Mastering Math for the Building Trades

Mastering Math for the Building Trades

James Gerhart

McGraw-Hill

New York San Francisco Washington, D.C. Auckland Bogotá
Caracas Lisbon London Madrid Mexico City Milan
Montreal New Delhi San Juan Singapore
Sydney Tokyo Toronto

McGraw-Hill

A Division of The ***McGraw-Hill*** *Companies*

The sponsoring editor for this book was Zoe G. Foundotos, the editing supervisor was Steven Melvin, and the production supervisor was Sherri Souffrance. It was set in Garamond through the services of the PRD Group.

Printed and bound by R. R. Donnelley & Sons Company.

McGraw-Hill books are available at special quantity discounts to use as premiums and sales promotions, or for use in corporate training programs. For more information, please write to the Director of Special Sales, McGraw-Hill, 11 West 19th Street, New York, NY 10011. Or contact your local bookstore.

6 7 8 9 10 DOC/DOC 0 9 8 7

ISBN-0-07-136023-9

Contents

Introduction

Mastering Math for the Building trades is intended to serve everyone involved in the various trades of the construction industry: contractors, tradespeople and suppliers, from the old pro to the newest person on the job. It will serve as a valuable reference tool in the office, at the counter, or on the job site. Used as a learning text for the new man on the job or for the pro who wants to expand his current abilities, this book is your guide to an easier understanding of the math encountered in the building trades.

Math, like English, is a language expressing specific ideas, theories and answers. The language of math is numbers. If we are well versed in this language of numbers, it will help us make important decisions and perform our tasks more effectively and efficiently.

This book gives examples of the various kinds of math used within the trades, while providing translations of the confusing language of math. Each chapter contains applied mathematics in relevant examples with each concept, equation and result explained in simple, easy to understand terms. Practice questions at the end of each chapter let the user become familiar with algebra, geometry and trigonometry to the point where they will no longer be foreign languages.

In this second edition, in-depth sections on such topics as: fluid mechanics, septic systems, fiber optics, metal framing and computer software have been added. More illustrations have been included throughout the book to help demonstrate the concepts and practices presented in the text. There are also many more tables included in this edition, making everyday numbers crunching easier while in the field.

With over 100 illustrations, and many valuable tables which simplify everyday calculations, this book walks the reader through essential mathematical equations. Using step-by-step procedures, each chapter demonstrates not only how to perform the math, but also why the math needs to be performed.

Chapter 1 explains how to use the tables presented throughout the book, and introduces the basic use of pocket calculators and personal computers to help solve basic building trade problems.

Chapter 2 not only introduces various measuring tools used by the construction trades but also discusses and provides examples in diverse areas of measurement. Some of these areas include: problem solving using steel squares, using different scales of measure, utilizing framing tables, the brace table and the Essex board measure table.

Chapter 3 introduces various techniques used in the planning and administrative aspects of building. Included are traditional and updated methods of performing many tasks. Concentrating on computer software, topics include: estimating, service and equipment management, labor productivity, time and material contracts, billing and purchasing. Other areas discussed include: HVAC load calculations, duct sizing and layout and energy analysis.

Chapter 4 deals with site preparation. The grading and excavation include topics including: measuring grade, plotting areas and the dimensions of excavation. Also included in this chapter is a useful discussion on finance. Tables breaking-down monthly payments into interest and principal, examples of compound interest and the use of financial calculators help you make informed financial decisions.

Chapter 5 covers both arithmetic and tabular estimating procedures for the many areas of the concrete and masonry trades. Methods of selecting mix proportions, estimating materials and labor are hot topics. Estimating brick for various patterns, thickness and joint size are discussed along with concrete block and natural stone.

Chapter 6 is the first of four chapters to examine the math involved in carpentry. Included here are the topics of floor framing and covering. Numerous tables highlight discussions of determining live and dead loads, beam selection, calculating column, post and joist size among other topics.

Chapter 7 deals specifically with the aspects of wall framing. Calculating the number of studs required, framing for windows and doors and partition framing are included. An introduction to metal framing is presented along with discussions on ceiling joists, exterior walls, insulating board sheathing, plywood and painting. Also included is a detailed section pertaining to the thermal performance of the envelope and its components.

Chapter 8 covers the calculations and required when constructing a variety of roof styles. Topics covered include roof rise and span, common roof layouts, the steel square's rafter table and deriving roof layouts. Detailed discussions on various layouts and complications are illustrated and rich with examples.

Chapter 9 details the math involved with estimating the materials and labor concerning the main areas of interior finishing, including stairs, walls, ceilings, floors, windows and doors, millwork and paints.

Chapter 10 presents the mathematics required when planning, laying out and troubleshooting heating and cooling systems. Topics discussed include: load calculations, thermal transmittance (U-values), shading coefficients of fenestration, wall and glazing areas, internal cooling loads and pick-up loads. Other topics of interest are: proper control zones, ventilation and air exchange.

Chapter 11 addresses the math needed for planning, estimating and installing water supply systems and their associated plumbing. Areas covered include: measuring water flow, friction loss, pipe runs and layouts, velocity head and equation coefficients. The components, design, layout and function of septic systems is also covered in this chapter.

Chapter 12 covers the mathematical formulas and estimating math used by electricians for determining usage needs and materials estimating. Basics, such as calculating voltage, amperage and resistance, wire types and sizes combine with new topics such as fiber optics, data transmission and computer networks.

Chapter 13 provides a summary of various construction math procedures as a quick reference.

As building methods and products continue to evolve, there is more and more crossover between trades. There is also more crossover between units of measure. The appendices in the back of this book provide easy to use references on these topics. The glossary includes commonly used terminology from each of the individual trades. A moment taken to insure that you and your fellow craftsmen are on the same page can save a lot of time. Appendix A includes units of measure most often used in the various building trades. This useful table provides conversions from one unit of measure to another while providing the commonly used abbreviations for each unit. More frequently today than ever measurements are mixed, not all are in fractions nor are they all in decimals. Appendix B deals solely with the conversion of fractions into decimals and decimals into fractions. The answers to the end-of-chapter questions are located in Appendix C. Working through each of these questions and checking the answers with Appendix C will sharpen your skills and make your job easier and more productive.

This book is intended to help everyone, even the numerically challenged, to understand the basic calculations and concepts of math used throughout the building trades. With this valuable aid as your companion, you too can be more productive. Learn the language of math and work smarter instead of harder. Use this book to improve your bottom line and make numbers serve your needs.

1

Math and the building trades

The easiest way for contractors to find themselves facing difficulties today is for them not to have a sound understanding of the mathematics involved in their business. From estimating quantities and costs on material and labor to keeping accurate financial records, to laying out the plans of a project, today's contractors implement a wide range of mathematical formulas on a daily basis. Not fully understanding these formulas or the appropriate uses for each can quickly take a profitable bottom line and turn it into a devastating loss.

Tables

Within each building trade, many repetitious calculations need to be made on each job to obtain the required information for different segments of the same job—for example, figuring all of the different header sizes for varying wall spans, or calculating space requirements for drainage lines.

In order to reduce the amount of time spent with such repetitious calculations, an abundance of tables are provided throughout the following chapters. Each table deals with a specific situation (such as the example above concerning headers and wall spans) presenting the results of the necessary calculation(s) for each of the most commonly occurring variables within the given situation. Following through on our example of determining header sizes for various wall spans, Table 1-1 illustrates how this information is presented.

Table 1-1 Header sizes for common wall spans.

Wall span, in feet	Header size, in inches
3½	2 × 6
5	2 × 8
6½	2 × 10
8	2 × 12

Finding the maximum span in the left-hand column of the table, simply read across to the right-hand column to find the appropriate-size header for that particular span. Each table in this book is similarly constructed for ease of use and to reduce the amount of time required for determining this information.

Prior to the presentation of each table, the equation(s) used for the calculation(s) within the table will be discussed and explained in terms of how they apply to that circumstance, enabling the contractor to deal with variables which may not be covered in the table.

Calculators and computers

Thanks to today's technology, performing complex calculations has become the realm of calculators and computers. The development of the silicon chip has enabled us to move from primitive counting methods and performing tedious calculations with pencil and paper to an age where hundreds of thousands of calculations are processed in moments. Along with the time savings afforded by these tools, another benefit is the consistent accuracy of the answers.

Calculators

The first electric calculators were bulky, expensive and only suited to desktop use, because they needed to be plugged into a wall outlet. They performed simple arithmetic functions (addition, subtraction, multiplication and division) with limited (if any) capacity for memory. As technology advanced, successive generations of calculators increased in capabilities while decreasing in size, although they were still rather large. Today calculators are manufactured to meet just about any specific need—business calculators that perform complex financial analysis or statistical calculators that can handle a variety of tasks dealing with sampling, averaging, and performing linear regressions as well as algebraic equations. Many calculators can handle functions necessary to two or more specific areas, such as business and statistics.

Each successive generation of calculators has included changes that make performing complex calculations easier. For example, most pocket calculators today have sufficient memory storage to allow several mathematical operations to be performed stored and the results used to perform yet another calculation. Finding the square root of a number today is as simple as entering the number and pressing one key on the calculator. One of the most significant advancements in calculator functions is the ability of many of today's calculators to accept algebraic equations entered as one non-stop algebraic statement, just as it would be written on paper. If the equation is $X = 3(A + B)$, simply press the keys in the following sequence:

Key: 3
Key: × (multiplication sign)
Key: Left parenthesis
Key: A (the number "A" represents in the equation)
Key: + (the addition sign)
Key: B (the number "B" represents in the equation)
Key: Right parenthesis
Key: = (the equals sign)

The resulting number in the display is the answer to the equation.

Scientific calculators are also available; these specialized calculators are designed for users who are involved in the sciences, engineering, electrical/electronic, and technical fields. What makes these calculators so unique is that many of them are programmable; in other words, in addition to the functions and programs built into the calculator at the factory, the machine can also be programmed by the user to perform alternate-situation and application-specific tasks as well.

The historic shortcoming of pocket calculators is that most do not provide hard copy (a paper printout) of the calculations performed on them, which makes spotting an error made during data entry practically impossible. For this reason, you should always reinvest some of the time saved by using these tools to run the calculations more than once in order to compare the results. Remember, the calculator will not make mistakes, but it is only as accurate as the numbers punched into it by the user.

Computers

For contractors, the days of "guesstimating" the cost of a job and a client's bill are long gone, as are the days of keeping records on stray pieces of paper (or for those who are organized, keeping records in actual journals), or trying to track labor hours and overtime costs by hand. The reason for these things being outdated is the coming of age of the computer.

For decades the computer was solely the province of NASA scientists, mathematicians, universities, large accounting firms, and other big businesses that could afford not only the computers themselves but the rooms and personnel which these machines required. Early computer systems were enormous machines requiring vast amounts of space completely separate from the rest of the work environment, in order to have the precise temperature and humidity conditions necessary in order to function properly. In addition to the space requirements, specialized personnel who spoke one or more computer languages were required to maintain, program, and operate the computers. Programming a computer to perform a simple calculation involved one of these specially educated and trained individuals to make a series of punch holes in a number of cards that were then fed into the computer—all in all, a time-consuming and expensive proposition.

Today, however, the computer has decreased in size so that it sits on a desk top or can be carried in your hand or briefcase (as in the case of "laptop" computers). Computers today no longer need special environmental conditions, nor do they require a trained technician for daily operation; they have reached a level of "user-friendliness" that allows their almost limitless power to be used by almost everyone, including contractors.

In a survey conducted by a leading supplier of software to the construction industry, the amount of money that could be saved by a typical construction company switching from a manual to an automated job costing and financial system was estimated to be in excess of $100,000.

Profiting from computerization

There are many major ways a construction company can profit from computerization, including controlling costs during each job phase, estimating costs more quickly and accurately, reducing workers' compensation by tracking work hours by risk category, organizing accounts payable for better use of available vendor discounts, speeding up and increasing the collection of accounts receivable, eliminating external service bureau costs, improving clerical productivity, and even more.

A variety of hardware (the machines themselves) and software (the programs that make the machines run) are available today, from very basic systems to very elaborate systems. Which computer and what programs an individual contractor will require will largely depend on the size of the company and how many tasks the computer will be expected to control. In recent years, the cost of a personal computer (regardless of manufacturer) has dramatically fallen, while at the same time the capabilities of these machines have drastically increased. For example, the average personal computer of today, which ranges in price from about $1200 to several times that amount (depending on how many bells and whistles are purchased with the system) is 20 times more powerful than the millions-costing computers that NASA used to send the first Apollo mission to the Moon. Do not be intimidated by this; it simply means that a personal computer is capable of handling most of the jobs that require number crunching, memory, or analysis.

Controlling costs during the job

One of the major problems for any contractor is being able to track, on a timely basis, the job costs of work in progress. Computerization solves this problem by keeping job cost data up-to-date, providing ample time for the contractor to react to cost overruns before they cut into profit margins. A generic program called a *spreadsheet* can be used for this function.

A spreadsheet is a numbers-crunching program, which can keep track of hundreds of different variables in its data bank, perform all the required calculations on those variables, and present a bottom-line answer. One of the versatile functions of spreadsheet programs is their ability to perform "what-if" analysis. This function allows the contractor to store all of the normal variables involved in keeping job costs under control and then change one of those variables to see what affect the change will have on the other variables and the final outcome. This is done by simply changing one of the stored variables; the computer then automatically recalculates all of the other factors and present the new final answer—all accomplished within a matter of seconds and without the chance of human error in the calculations. Aside from the generic spreadsheet programs, there are specialized programs created for the construction industry that incorporate spreadsheet functions along with many other software program functions.

Estimating more accurately

Today, success depends on how close to the mark a contractor can come in estimating costs and coming up with realistic bids that will win contracts and still be profitable. Whether it's bidding, negotiating, planning, budgeting, value analysis, pricing change orders, or evaluating quotations, however, good estimating is the key to profitability. In the past, most estimators have had only rulers, calculators and pencils to assist them. Now, the computer has recently proven itself to be valuable in estimating by allowing for more bids of higher-quality estimates (reflecting standard company policies) to be generated in much less time than performing these tasks by hand.

A computer and the software that runs on it are tools that can help the contractor understand the uncertainty in a bid. Just like a chainsaw cutting wood faster than a handheld saw, a computer can help analyze information quicker and easier than by using a pencil, paper, and calculator. Each bid could contain hundreds of variables, and any one factor could make the difference between winning and losing a bid, or the resulting profit or loss. Once again, the spreadsheet program is a major player in preparing bids, and its primary function is to calculate the final bid price. As is done on paper, the spreadsheet can list items, apply quantities and unit costs, include overheads and profit, and then add the numbers to get a price. This program can keep track of almost anything, including competitors' markup margins and pricing.

Reduced workers' compensation insurance costs

With a manual system, workers' compensation insurance costs are usually calculated by multiplying the total number of hours worked by the highest premium rate that applies to that type of work. With a computerized system, a contractor can break down the hours worked into risk categories, and base insurance premiums on the various categories of work performed. The result could be a substantial reduction of insurance costs.

Take advantage of vendor discounts

In many accounting systems, the accounts payable module (program) offers a feature that enables a contractor to take full advantage of available vendor discounts. By inputting due dates and discount terms for each vendor into the system, the computer can automatically cut checks to the individual vendors within the discount time frame in order to realize these savings. One of the many benefits to the contractor from a computerized system is that reports can be generated at any time. For example, reports showing which vendor discounts were taken this month and which were not can be reviewed by the contractor in order to fine tune the company's financial picture.

Invoice customers on time

During peak construction periods it is not unusual to fall behind on paperwork. With a computerized system, contractors can utilize the capability of computer programs to integrate with each other to make paperwork a simple chore. Integration is the ability of one software program to "speak" or "communicate" with another software program. For example, when a job is finished, an integrated financial package can be programmed to take the appropriate information from the spreadsheet program and generate an invoice, usually with just a few simple commands on the keyboard.

Keep track of accounts receivable

Cash flow can be significantly improved by keeping a tight leash on accounts receivable. The same financial program mentioned earlier can also produce aging reports, categorized by customer and giving a receivable balance for each client who has not yet sent payment.

Using this tool, the contractor can respond quickly to delinquent accounts by assessing late charges or sending "reminder notices." Again, by utilizing integrated software, a word processing package can be programmed to automatically take the information from the financial program and produce a form letter to be sent to all customers who are presently 30 days (or any amount of time determined by the company) or more past due on their bills.

Eliminate service bureau costs

With the installation of a computer system, contractors no longer have the need to hire external service bureaus to process payroll or perform other accounting functions. Financial programs are fully capable of tracking employee hours (straight time, overtime, double time, holiday pay, etc.) and generating paychecks. At the same time, these same programs will keep track of bank balances, loan payments, and interest paid or earned. At tax time, reports can be generated summarizing or detailing by item all revenues and expenses categorized by job, by employee, by vendor, by loan number or name, or any other criteria selected, making income tax preparation much easier and less costly.

Eliminate errors

As long as the proper figures are entered into the computer system to start with, any reports, analysis, bids, invoices, etc. coming out of the computer will be accurate. Remember the old saying among computer users: "garbage in = garbage out." All in all, computers are much more accurate than humans when performing everyday calculations. Some of the errors that can be avoided by using a computer are as follows:

- Billing errors that can result in under-billing or giving customers an excuse to delay payment.
- Incorrectly calculated payroll taxes that can result in penalties from government agencies.
- Accounting errors which tend to increase audit time and costs.
- Mistakenly paying invoices twice.

With an integrated computer accounting system, any errors made during data entry only need to be corrected once, all relevant accounts, reports and records are automatically changed and the data corrected.

Increase and improve clerical productivity

Computer systems free up clerical workers so they have time to tackle more productive tasks, such as generating management reports. The quality of the work is higher, mistakes are fewer, morale is higher, and in most cases the number of clerical employees required is fewer, thus reducing overhead.

Effectively manage finances

Computerized systems make it easy to produce up-to-date reports on accounts payable and accounts receivable as well as being able to generate complete financial statements anytime of the year (not just at year end). These reports are very helpful when dealing with financial institutions, since they present information about the contractor's business in the same language that the bank speaks. Bankers tend to have more confidence in computer prepared statements than in manually prepared statements; and when detailed, organized, complete financial information is part of the contractor's loan presentation, it reflects well on his/her professionalism.

Smart decisions

A major productivity feature of advanced software systems is the "work package." A work package allows a contractor to take off a job's numerous items just by entering the appropriate information. In the instance of a roofing contractor, to take off a 4-ply built-up roof, for example, enter the roof's length, width, and slope. The work package will automatically estimate labor, material and equipment costs for all the felts, asphalt, surfaces, fasteners, and any other items related to that roof. Based on the length of gravel stop, the intelligent work package will count the number of joints required, add the proper amount of material for the lap, and round up to the next 10-foot piece.

From this take-off, the work package will calculate estimated labor and material costs for numerous items, including insulation, fasteners, felts and surfaces, as well as taking into account waste factors and transferring information into a complete bill of materials and a detailed field productivity report.

Reduce overhead

Because today's computer programs have been designed to be "user friendly," the time it takes to learn how to use them is relatively little. An employee who has little or no formal accounting experience or education can learn how to enter the financial information into the accounting program at the right places and then the computer does the rest. Similarly, word processing packages are very simple to use, allowing office staff to handle correspondence faster and more accurately with fewer labor hours. Finally, the intelligent work packages will allow contractors to be more efficient in preparing bids, generating more bids in less time, which will result in winning more bids, and freeing up the contractor for other tasks.

2

Measuring tools

Before harnessing the power of calculators and computers to solve complex equations, contractors must still measure each phase of the job they are involved with in order to obtain accurate numbers, which are "plugged" into those equations. The tools discussed in this chapter provide the means for accurately obtaining the necessary figures in most situations.

Rules and tapes

Different forms of rules and measuring tapes have been employed by the construction industry for centuries—from simple sticks with markings sawn into them to highly specialized instruments with multiple scales of measure accurately engraved into them. Some of the most useful of these tools are discussed in this chapter.

Measuring tapes

Available in many different lengths and widths, these measuring tools can be used for distances ranging from a fraction of an inch to several hundred feet. The shortest of the tapes (usually no longer than 12 inches) roll into a coil for storage, but when straightened out have a curved crown that maintains the rigidity of the tape while in use. The edges of the tape will remain flat and straight and should be placed vertically along the surface to be measured.

As with steel rules, tape measures usually afford more than one scale of graduation. The more common lengths of measuring tapes roll up into a coil housed in a metal or plastic case (Fig. 2-1). Tape measures come in various sizes and shapes from pocket-sized to the long tape measures as pictured in Fig. 2-2. A hook on the free end of the tape is used to secure the tape onto or immediately adjacent to the surface to be measured. This hook makes up a portion of the first unit of measure; the metal hook on the end is counted as part of the first inch of measure, so the hook should not be wrapped around a corner, unless that corner is to be part of the overall measurement.

Tape measures must be supported the entire length of measure in order to avoid any sagging, which would cause an error in the reading. Securing the hook to the starting point of the measure and pulling the tape taut will sufficiently reduce sag, but always remember to sight down the length of the tape to make sure that the plane of the tape is level and not inclining or declining, which would again result in a false measurement.

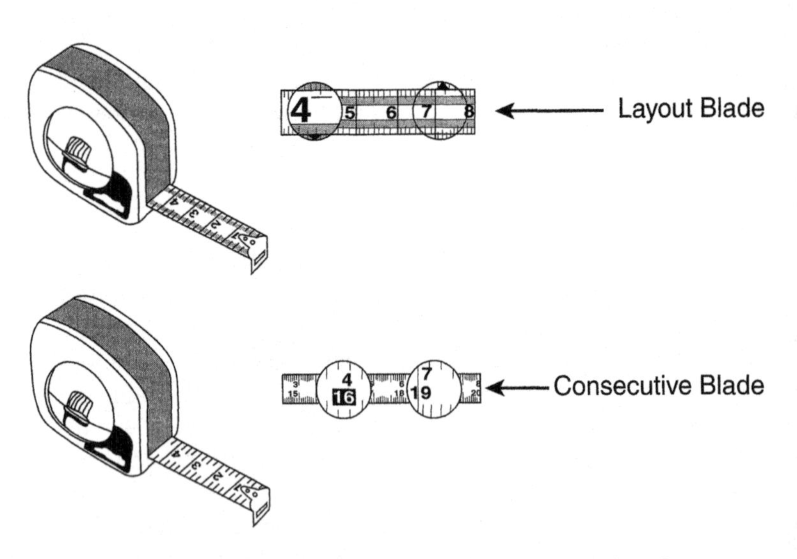

2-1 *Tape measures are available with various scales of graduation. Courtesy of Cooper Tools.*

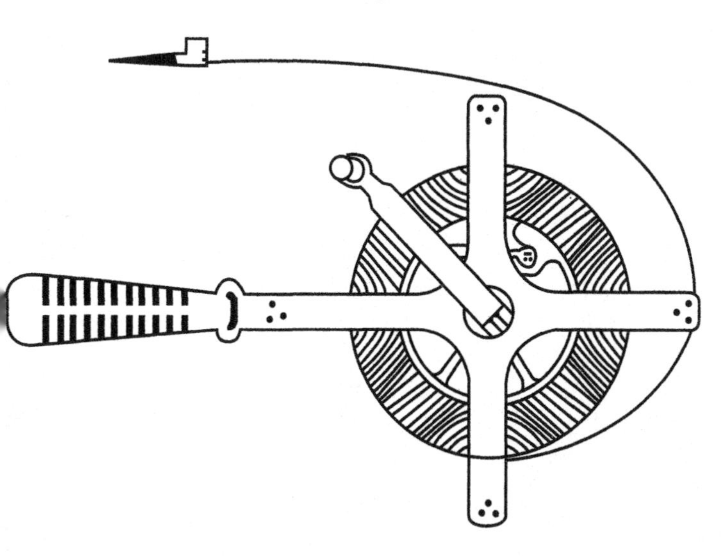

2-2 *Long measuring tapes allow greater versatility. Courtesy of Cooper Tools.*

Measuring wheels

The distance measuring wheel is a real time-saver for all the trades. Wheels automatically record feet and inches line-to-line, wall-to-wall, vertically, overhead, around curves and over contours of any firm surface.

Measuring wheels come in a variety of sizes and shapes as seen in Fig. 2-3. To operate, squeeze the trigger to release the brake. Set the disc to zero. Place the wheel at the starting point. Squeeze the trigger and begin wheeling off the distance required. If you need to stop, reset the brake by squeezing the trigger. To check the distance covered, read the number of feet on the counter and inches on the disc.

Electronic counters add a new level of performance and versatility to measuring wheels. An easy-to-read LCD display shows distances either in feet or in meters simply with the push of a button. Options include dual counters that measure total distance or segments. Some measuring wheels operate in forward and reverse. Two counters operate in either direction, adding or subtracting accumulated distance.

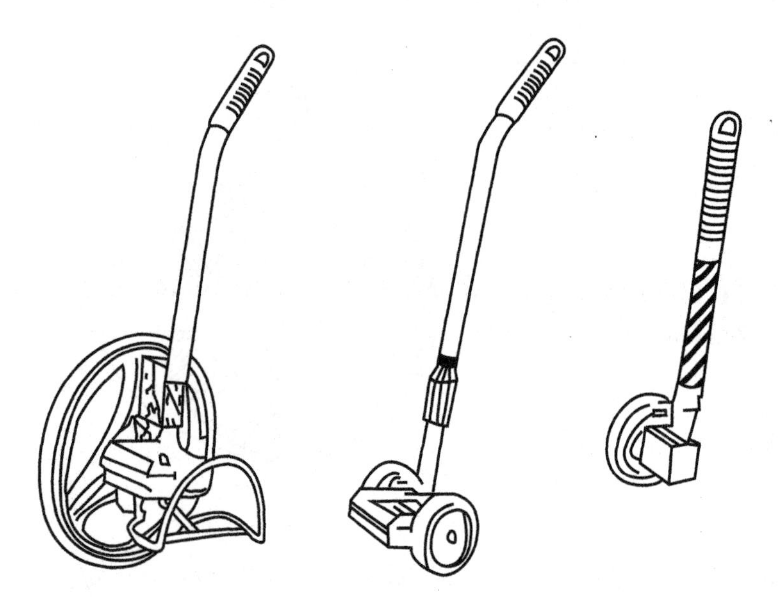

2-3 *Typical measuring wheels. Courtesy of Cooper Tools.*

Folding rules

Although useful for rough measurements, folding rules (as seen in Fig. 2-4) are not as precise as other instruments because of the folding hinges. These rules are available in different sizes (usually from 2 to 6 feet in length) with each 12-inch segment hinged to fold against the next section. Over time,

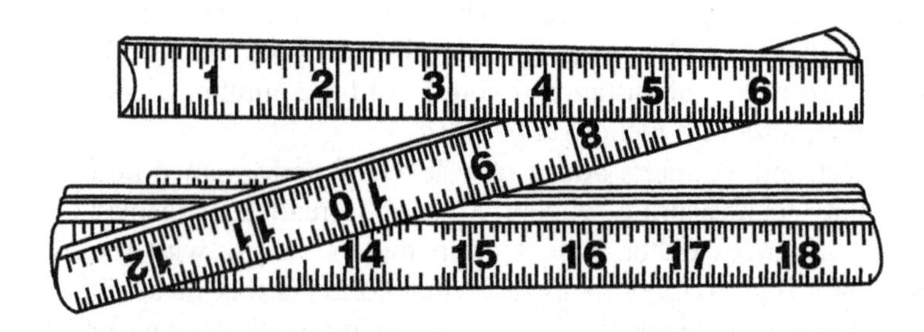

2-4 *Folding rules have been used for generations. Courtesy of Cooper Tools.*

these hinges become loose-fitting and will often result in a sagging or peaking of the rule, which throws off the accuracy of the measurement.

Steel rules

The most simple of measuring tools, the steel rule is still one of the most widely used. Although made of plastic, steel, or fiberglass, this tool is still most commonly referred to as a steel rule and is available in a variety of lengths, the most common of which are 6 and 12 inches.

Along each of the four edges of the rule (front top and bottom, back top and bottom) is a graduated scale of measurement separating each inch into equal portions. Each of these scales employ a different scale resulting in different graduations (marks identifying units of measure). For example, on one side, one edge may divide each inch into $\frac{1}{8}$-inch segments, while the other edge separates each inch into $\frac{1}{16}$-inch segments. Usually on the opposite side of the rule one edge will divide each inch into $\frac{1}{32}$-inch segments and the opposite edge will divide each inch into $\frac{1}{64}$-inch segments.

By starting at the far left side of the rule and placing the edge of the rule onto the surface to be measured, you can receive the most accurate measurement. Then, count to the right any number of inches or fractions of an inch to determine the needed measurement. Remember to always reduce fractions to their lowest forms. For example, dividing $\frac{4}{16}$ by the largest whole number which goes into each portion of the fraction (numerator/top and denominator/bottom) equally results in the fraction reducing to $\frac{1}{4}$ ($\frac{4}{4}$ = 1 and $\frac{16}{4}$ = 4). Normally, the $\frac{1}{4}$-inch and $\frac{1}{2}$-inch graduation lines on each scale are made longer than the lines denoting the smaller graduations, in order to facilitate easier counting.

Many different variations of the steel rule are available. Varying thicknesses, widths, lengths and graduations are produced to meet just about any situation. Thinner rules are more flexible and can fit into tight quarters. Some rules are designed with scales across the short ends of the rule to make measuring in small areas easier. Many of today's rules also include at least one edge with metric graduations, which is becoming increasingly necessary on jobs where building materials are manufactured in metric sizes.

Story poles

Brick and masonry contractors have long used story poles as a short cut method to ensure that each course of material is at the proper height. Normally a good straight 2 × 4 with a series of marks clearly made on all four sides, set at precisely measured regular intervals, this tool allows the tradesperson to quickly and accurately check the courses of work to ensure the uniform progress of the job.

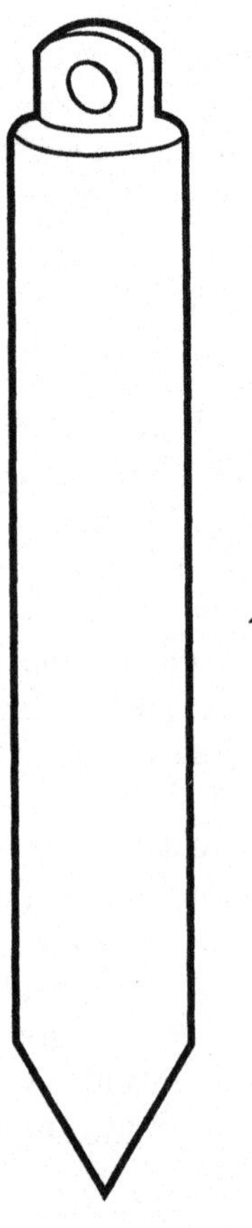

2-5
Typical plumb bob.

Batter boards

A batter board is a board frame supported by stakes set back from the corners of a structure, which allows for the relocation of certain points after excavation. Saw kerfs in the boards indicate the location of the edges of the footings and the structure being built.

Plumb bob

A plumb bob (Fig. 2-5) is a weight connected to a plumb line, which is lowered to an elevation below the person using it, stretching the plumb line to make a straight line in order to align vertical points.

Steel square

One of the most useful, versatile, and easy to use tools is the steel square (sometimes referred to as a carpenter's square or framing square). The two portions of the steel square (the tongue and the blade) form a right angle (i.e., an angle measuring 90°) at their intersection. Connecting the blade and tongue along a third straight line forms a right triangle, the geometric principle upon which this instrument is based (Fig. 2-6).

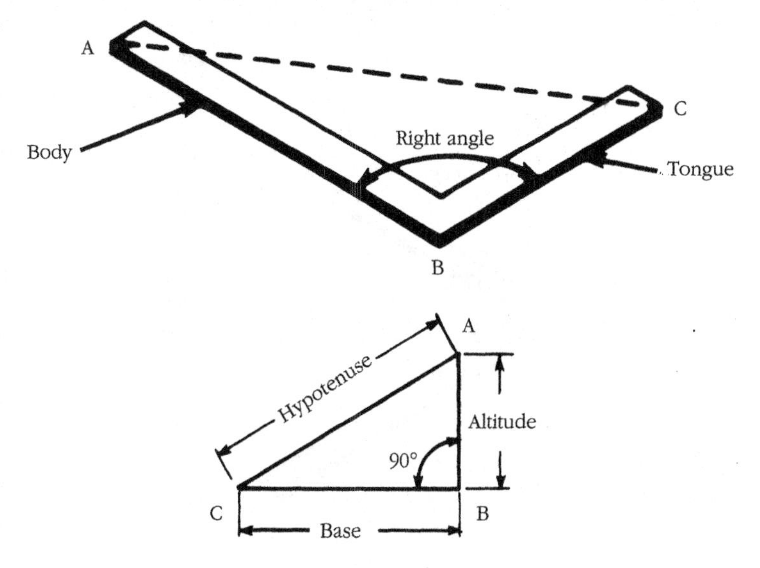

2-6 *The tongue and blade of the steel square form a 90-degree angle.*

Steel squares come in a variety of sizes, but the most common usually have a 2-foot long by 2-inch wide blade with a 1½ foot long by 1½ inch wide tongue. The steel square is equipped with eight graduated scales—one on each edge of the face and back of the tongue and one on each edge of the face and back of the blade (Fig. 2-7). Each of these scales subdivide each inch by different units of measure. The scales located on the face of the square (normally the side with the manufacturers name stamped into it) subdivide inches into common measures of eighths of an inch and sixteenths of an inch. Graduated scales of tenths and twelfths are found on the back side of the steel square.

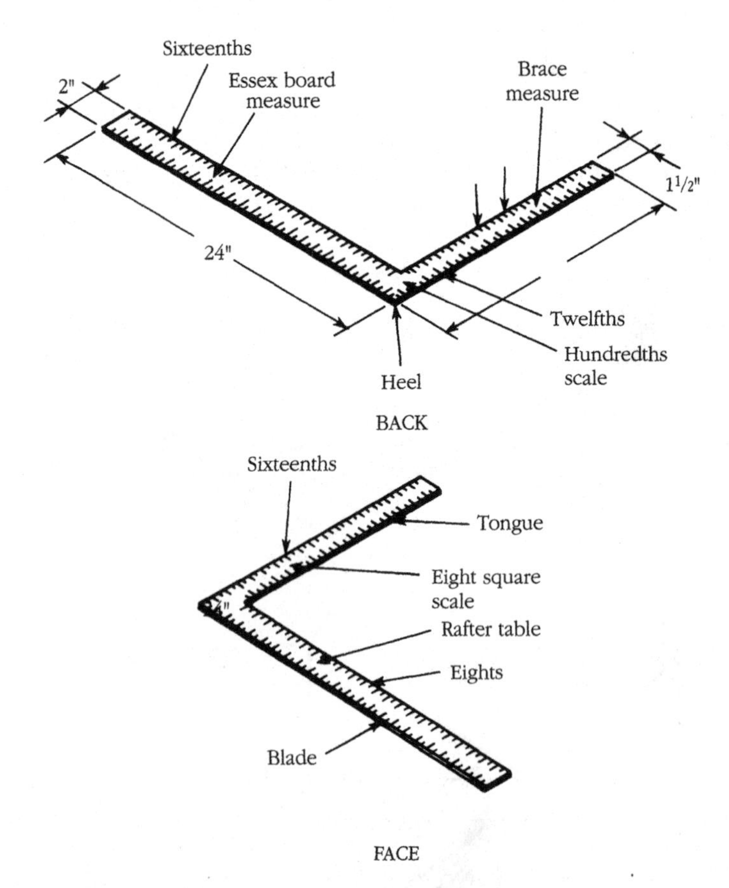

2-7 *The graduated scales of a steel square.*

Problem-solving using steel squares

When the problem involves determining the length of a roof rafter, for instance, view the elements of the roof geometrically. Visualize the rafter in place, combined with the other two elements of the roof framing to form a triangle. The question really being asked, then, is "What is the length of the hypotenuse of this right triangle?" (The hypotenuse is the longest leg of the triangle formed by the rafter.) The lengths of the two shorter legs of the triangle are already known. The rafter (or the member forming the longest leg) can also be called the hypotenuse of the right triangle (Fig. 2-8).

The first step in solving this problem and others like it is to reduce the dimensions to a size or proportion that will fit inside the steel square. This is accomplished with the use of ratios. First, organize the known measurements. For this example, assume that the total rise (the vertical side) and the total run (the horizontal side) are each 60 inches long and remember that the standard unit of run is always taken as 12 inches and is measured on the tongue of the steel square. The "unknown" factor is the unit of rise. The unit of rise can be expressed as X in the statement "60 is to 60 as 12 is to X." This ratio is more concisely written mathematically as $60 : 60 :: 12 : X$.

The next step is to convert the ratio into a workable equation, as follows:

$$\frac{\text{Total run}}{\text{Total rise}} = \frac{\text{Unit of run}}{\text{Unit of rise}}, \text{ or}$$

$$\frac{60}{60} = \frac{12}{X}$$

The problem has now been reduced to a workable algebraic equation. When dealing with algebraic equations, the goal is to isolate the unknown factor (in this case the X) on one side of the equal-to sign (=), with all the known or calculable values on the other side. The first step in achieving this result is to simplify the equation; the term on the left of the equation 60 / 60 can be reduced by performing the math of dividing 60 by 60, which equals 1; so the equation can be rewritten as

$$1 = \frac{12}{X}$$

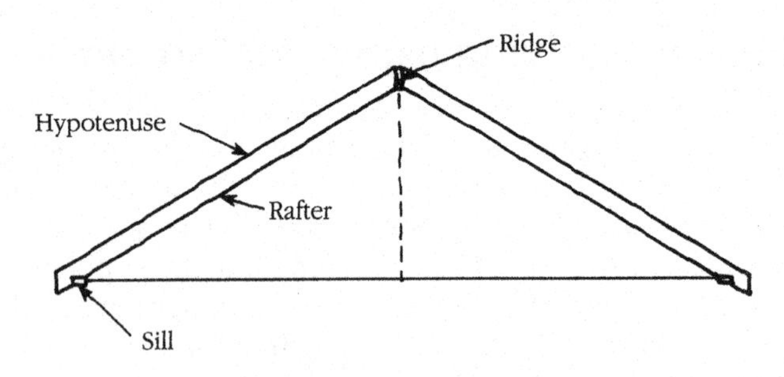

2-8 *Roof-framing members form a right triangle, with the rafter forming the hypotenuse.*

Because the object is to isolate the unknown value (X) on one side of the equation, the next step is to eliminate X from the right side of the equation. Currently the equation calls for 12 to be divided by X, if both sides of the equation are multiplied by X the new form of the equation would be

$$1X = 12$$

$1X$ can also be written as just X, so the equation is now

$$X = 12$$

The unknown value (X) has been isolated on one side of the equation, and in the process the value of X has been established. The remaining statement ($X = 12$) is the answer to the equation, which means 12 is the unit of rise.

Quick hint: Any time the total run and total rise have the same value (measurement), the unit of run and unit of rise will also have the same value.

In situations where the total run and total rise are not the same value, simply work through the equation. For example, assume a total run of 84 inches and a total rise of 96 inches; the unit of run remains 12 inches because, as stated earlier, it's a constant. The statement $84 : 96 :: 12 : X$ is the ratio expressing this situation, with X representing the unit of rise. The equation would be

$$\frac{84}{96} = \frac{12}{X}$$

Again, to isolate X on one side of the equation, multiply both sides by X. Multiplication is used because currently the equation calls for 12 to be divided by X, so multiplying by X will eliminate X from that side of the equation. The resulting equation reads as follows:

$$\frac{84X}{96} = 12$$

In order to isolate dissimilar terms on different sides of the equals sign, multiply both sides of the equation by 96 to get rid of the 96 divisor on the equation's left side. This results in the equation:

$$84X = 12 \times 96, \text{ or } 84X = 1152$$

By dividing both sides of the equation by 84 (again to reduce the equation to simplest terms), the resulting equation is

$$X = 13.7$$

X therefore represents 13.7, the unit of rise. For practical purposes, 13.7 can be approximated to 13¾ inches on the job. (See Appendix B for conversions from decimal to fractional.)

Line length

Once the unit of rise is established, set the unit of run on the tongue of the steel square and the unit of rise on the body of the steel square. After setting the steel square on the edge of a board (having a true, planned, straight edge to it) and lining up the appropriate marks on the tongue and body of the steel square, draw lines down from these marks onto the board. The measurement between these two marks is the length of the triangle's hypotenuse.

In situations concerning a right triangle, where the base of the triangle is equal to the total run and where another side (the altitude) is equal to the total rise, the measurement of the

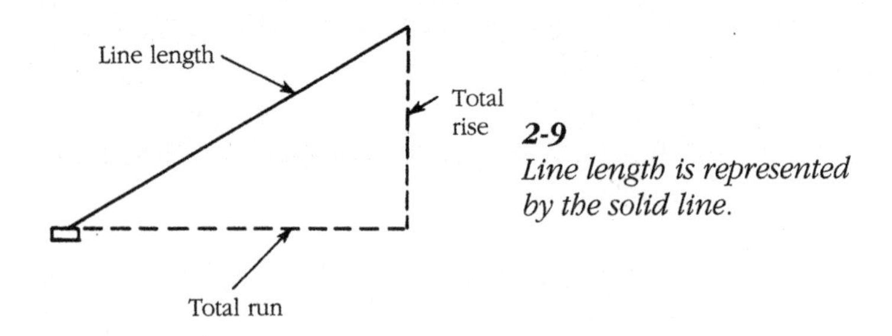

2-9
Line length is represented by the solid line.

distance between the two drawn lines is also known as the line length (Fig. 2-9).

By multiplying the bridge measure by the number of times the total run is divisible by the unit of run, the total length of the rafter is determined.

Using the twelfths scale

As mentioned earlier, the steel square offers a variety of graduated scales, one of which is the twelfths scale, which divides each inch into twelve equal segments. This scale of twelfths can simplify proportional reductions; by allowing each one twelfth section (graduation) to represent one inch instead of $\frac{1}{12}$ of an inch, reducing the size of a triangle becomes much less complicated.

By employing the twelfths scale and allowing one inch to equal one foot, an actual measurement of 5 feet 6 inches can be quickly reduced to 5 and $\frac{6}{12}$ (or $5\frac{1}{2}$) inches on the twelfths scale. Of course the reverse is also true; using the steel square to figure the length of the roof rafter (as discussed earlier in this chapter) results in a measurement of $10\frac{3}{12}$ inches on the twelfths scale, which simply converts to 10 feet 3 inches of actual rafter length.

The tenths scale

Another alternate scale of measure located on the steel square is the tenths scale. This scale is graduated in tenths of an inch and usually found along the inner edge of the back of the tongue (Fig. 2-10). This scale, used in conjunction with the sixteenths scale found on the inner back edge of the body of the square can also perform proportional functions.

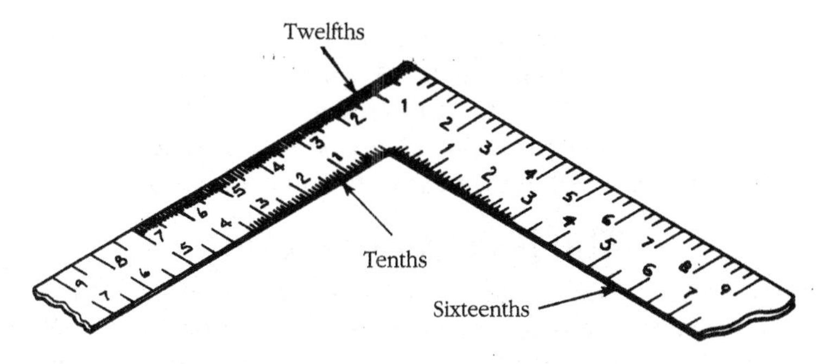

2-10 *The tenths scale, found on a steel square.*

Assume that 60 feet of excavation has taken eight hours; also assume that there is a need to know how many feet of excavation can be accomplished by the same crew in five hours. Use the steel square as follows:

1. Place the tongue of the square along a true, planed, straight edge on a board with the tenths and sixteenths scales in view. Make the assumption that each tenth of an inch is really one foot. Make a mark on the board which coincides with the 6-inch mark on the tongue scale ($6 \times 10 = 60$) to represent 60 feet.
2. Follow down the body of the square to find the 8-inch mark on the scale (to represent 8 hours of labor) and make a mark on the board coinciding with it.
3. Slide the body of the square down until the 5-inch mark lines up with the mark made on the board representing 8 hours.
4. Then, holding the square in place, use a straightedge and connect the mark made representing 60 feet to the current position of the 5-inch mark.
5. Where the line passes through the tongue of the square the reading will be 37$\frac{5}{10}$, or 37½ feet, which is the answer to the problem.

A faster method of performing this same proportional problem is with the use of a pocket calculator. Simply multiply 8 (the number of hours) by 60 (the number of minutes in each hour) which equals 480 minutes. Next divide 60 feet of excavation by 480 minutes to determine that 0.125 feet of excavation were accomplished per minute. The last step is to multiply 0.125 feet of excavation per minute by 300 (5 hours × 60 minutes) which will equal 37.5 or 37½ feet of excavation in a 5-hour period.

The hundredths scale

On the back of the tongue of the steel square, located in the corner near the brace table a scale graduated in hundredths of an inch is usually found. The placement of this scale is not random. Since the scale directly below it on the edge of the tongue is the twelfths scale measurements can be eyeballed between the two scales in order to easily obtain fractional measurements.

The hundredths scale, as the name implies, separates each inch into one hundred equal sections, with longer lines indicating twentieths and tenths of an inch (Fig. 2-11). This arrangement can be very helpful in computing rafter lengths with figures obtained from the rafter tables (presented in hundredths) discussed later in this chapter.

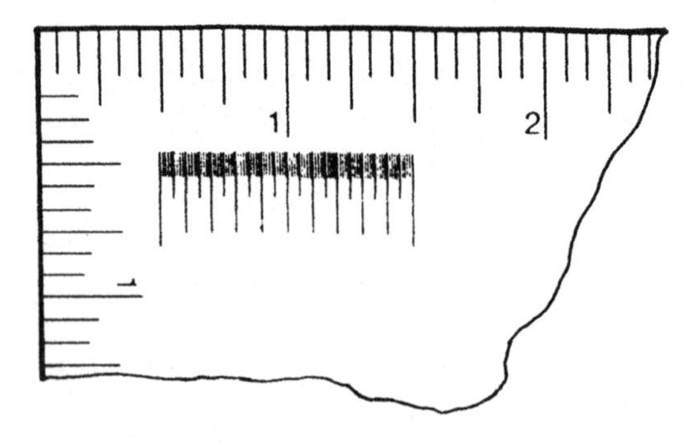

2-11 *The hundredths scale, found on a steel square.*

The octagon scale

Sometimes referred to as the eight-square scale, this is another graduation likely to be found on a steel square. Normally located at the middle of the face of the tongue, it is used to lay out eight-sided figures within squares of given dimensions (each being of an even number of inches).

Utilizing the framing tables of a steel square

In most cases a steel square will provide three tables on its surfaces for use in determining different construction requirements:

- The brace table: Located on the back side of the tongue of the square.
- The Essex board measure table: Located on the back side of the body of the square.
- The unit length rafter table: Located on the face side of the body of the square.

The use of the unit length rafter table will be postponed until it is discussed in depth in Chapter 9 of this book.

The brace table

As can be seen in Fig. 2-12, the brace table presents information about a series of right triangles, each having a run and rise of equal measure beginning with 24/24 and progressing to 60/60. Adjacent to these figures are the accompanying brace (hypotenuse) measures for each. This table cannot only be used to quickly determine the brace measure of a right triangle whose equal side measures are present in the table, but simple arithmetic allows the table to be used to determine the brace measure for almost any right triangle with two equal sides whose measurements are an even number.

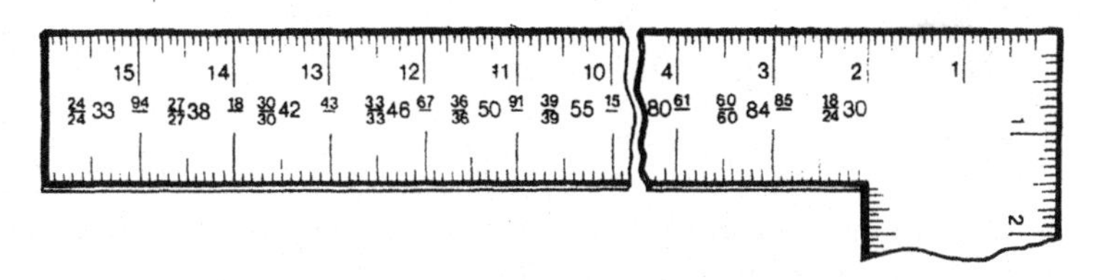

2-12 A convenient source of information, the brace table is located on the steel square.

For example, the first set of figures on the brace scale are for a triangle with two sides each measuring 24 inches and that has a hypotenuse (bridge measure) of 33.94 inches. If a situation arises that requires the bridge measurement to be determined for a triangle with equal sides of 16 inches, simply use ⅔ of the bridge measure above (22.63). Since 16 is equal to ⅔ of 24, it stands to reason that the bridge measures will also change proportionally.

Basic geometric principles state that any right triangle with two equal sides will be similar to any other right triangle with two equal sides. What this means is that the formula used earlier in this chapter can be used to determine the bridge measure for any similar triangle. For example, assume that a right triangle with equal sides of 40 inches is the subject for which the brace measure needs to be identified. On the brace table, find a set of equal side measures that are similar to the target triangle's measure. In this case, because 40 is a multiple of 10 and 30 is also a multiple of 10, 40 can be used as a similar set of measures.

Because the bridge measure for a right triangle with equal sides of 30 inches is known to be 42.43, an appropriate proportional statement would read 30 : 42.43 :: 40 : X, where X = the bridge measure for a right triangle with equal sides measuring 40 inches. Now, perform the same operations as before:

$$30 : 42.43 :: 40 : X$$

Multiply both sides by X and simplify:

$$30X = 40 \times 42.43$$

Simplify again to:

$$30X = 1697.20$$

Divide by 30 and get the answer:

$$X = 56.57 \text{ inches}$$

Quick hint: The most commonly found right triangle with unequal sides has proportions of 18 : 24 : 30 and is most often referred to as the "3 : 4 : 5 right triangle." Any triangle with these proportions must be a right triangle.

The Essex board measure table

Designed as a guide in figuring the number of board feet in a piece of lumber, the table is based on a standard 1-inch thickness for each piece of lumber. All calculations on the table begin with the column headed by the 12-inch graduation, which represents a piece of lumber 12 inches wide; for this reason, this column will be referred to as the starter column. The numbers in the column under the 12-inch graduation represent different lengths of lumber. As can be seen in Fig. 2-13, these numbers allow for calculations of lengths from 8-15 feet, with the column header (width of lumber) also serving as the 12-foot length figure.

To utilize this table, read down the starter column until the length of the piece of lumber in question is found. For example, assume a 10 × 7 × 1 piece of lumber is in question; begin by reading down the starter column to the number 10, then read left along that row until reaching a figure in the column under the number 7 (representing the width of piece of lumber in question). The figure at the intersection of that column and row (5.10) is the number of board feet (given in feet and twelfths of board feet) of the piece of stock in question, assuming the piece of stock is 1 inch thick.

For instances where the piece of stock in question is thicker than 1 inch, follow the same procedure as above and multiply the answer by the actual thickness of the lumber. For example, if the piece of lumber in the example above were 3 inches thick instead of 1 inch, the answer of 5.10 board feet would be multiplied by 3, resulting in a figure of 15.3 board feet.

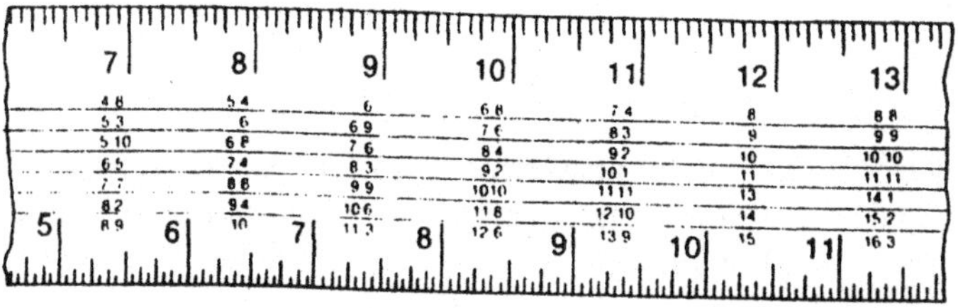

2-13 *Calculating board feet is made easy with the help of the Essex board table of the steel square.*

In the same manner, if a piece of lumber is longer than 15 feet (the longest tabulated length in the table) multiplication can be used with the table to obtain the required results. For this purpose assume a 20-foot by 8-inch piece of lumber. Read down the starter column to 10, then read across that row until intersecting the number within the 8 inch column (6.8 in this case). 6.8 board feet is the answer for a 10-foot by 8-inch piece of stock, so simply multiply by 2 for a 20-foot by 8-inch piece ($6.8 \times 2 = 13.6$).

In the case where the piece of stock in question is not evenly divisible by one of the tabulated lengths, simply separate it into two tabulated lengths. For example, to find the number of board feet for a piece of stock measuring 29 feet by 8 inch by 1 inch, start exactly as above to find the solution for a piece of stock 20 by 8 by 1, which equals 13.6. Now read down the start column to the 9-foot mark and read across to the 8-inch column to the answer: 6 board feet. Adding these two results together will reveal that a piece of stock 29 feet by 8 inches by 1 inch will yield 19.6 board feet.

Naturally, a calculator may be used to perform this function as well. A simple arithmetic formula correctly entered into a pocket calculator will determine the yield of board feet for any pieces of lumber:

The number of pieces of lumber × the thickness of the lumber in inches × the width of the pieces in inches × (the length of the pieces / 12)

This is where the *parenthesis* keys on a calculator (and the capabilities of the calculator to perform algebraic functions) come into play. Assume one piece of stock 21 feet by 9 inches by 1 inch. The data entry on the calculator would be:

$$1 \times 1 \times 9 \times (21 \div 12)$$

which will result in an answer of 15.75, or 15¾ board feet.

Quick hint: If all measurements are given in inches, divide by 144 instead of 12.

Table 2-1 provides a rule-of-thumb guide for figuring board foot measurements.

Table 2-1 Guide for board foot measurements.

Width (inches)	Thickness (inches)	Board feet
3	1 or less	¼ of the length
4	1 or less	1⅓ of the length
6	1 or less	½ of the length
9	1 or less	¾ of the length
12	1 or less	same as length
15	1 or less	1¼ the length

The surveyor's transit

The surveyor's transit is a precision instrument used in a process known as *differential leveling.* Using line-of-sight extends the transits imaginary plane outward from its own elevation. Swiveling 360° and tilting up and down allows for easy comparison of various points of elevation in different directions. Range and accuracy of measurement are greatly increased through the use of telescopic lenses in the transit, and horizontal crosshairs etched into its optics enhance the transits ability to determine the elevation of different objects at various locations. A detailed discussion of the role the transit plays in excavation can be found in Chapter 4.

Review questions

1. Add ½ and ⅓: (½ + ⅓)
2. Subtract ⅓ from ½: (½ − ⅓)
3. How many ⅓'s are there in 2?
4. Divide ½ by ⅓: (½ ÷ ⅓)
5. Find the value of x in the expression $16 : 8 = x : 5$

3

Basic math for estimating

This chapter explains the basics needed for preparing job estimates. However, note that the material and equipment prices and labor costs will vary with time and physical location. Therefore, any variables (material, equipment and labor expense) must be appropriately adjusted prior to fitting them into any equation.

Bill of materials

Estimating the materials and costs of a job is arguably one of the most math-intensive segments of any construction project. A list of all the materials required to complete the project (called a *bill of materials*) must be drawn up. The more precise the bill of materials is, the easier the project will progress, with less headaches and a fatter bottom line.

In many circumstances, the bill of materials will be drawn up by the draftsperson, based on take-offs and estimates, when the draftsperson prepares the original drawings. However, more and more contractors are taking on the responsibility of performing this function for themselves in order to assert more control over costs.

Materials take-off list

As the name implies, this document is a list of materials "taken off" of the plans for the project. The materials take-off list is an exact arithmetic accounting of each and every item specified on the drawings, including the name of each item, its description, the quantity required of each item and (if known) the part number, manufacturer, and stock number. A seemingly mundane job, this task is one of the key operations of any project. Without the correct materials on the job at the time they are needed, the project grinds to a halt.

Arithmetic method

Normally, a materials take-off list will begin listing those building elements required at the bottom of the project and move upward. For an example, see Table 3-1—a take-off list for a 40-foot building.

Notice that each item on the list in Table 3-1 is laid out in specific order:

1. *Item number*
2. *Item*: The name of the item.
3. *Number of pieces*: Assuming each footer requires three pieces, multiplied by 15 footers, equals 45 pieces needed.
4. *Unit of measure*: Common unit of measure for the item.
5. *Length-in-place*: The actual length of each member after preparation and prior to nailing.
6. *Size*: The quoted lumber size (2×4, 2×6 etc.).
7. *Length*: The commercial length of the lumber (8', 10', etc.).
8. *Number of pieces per length*: Refers to the number of times the length in place measurement can be divided into the commercial length of the lumber.
9. *Quantity*: The number of pieces of commercial length lumber required to supply the total number of pieces of length in place members.

When organizing a bill of materials, one must try to maximize the ordered materials.

For example, the section of the table headed as "number of pieces per length" must contain figures based upon the most economical use of each length of lumber. Using item number 1 as a reference, each footer has a length in place of 1 foot 5

Table 3-1 Example of a materials takeoff list.

#	Item	No. of pcs	Unit of measure	Length in place	Size	Length	No. of pieces per length	Quantity
1	Footer	45	Piece	1'-5"	2×6	10'	7	7
2	Spreader	30	Piece	1'-4"	2×6	8'	6	5
3	Foundation post	15	Piece	3'-0"	6×6	12'	4	4
4	Scabs	20	Piece	1'-0"	1×6	8'	8	3
5	Girders	36	Piece	10'-0"	2×6	10'	1	36
6	Joists	46	Piece	10'-0"	2×6	10'	1	46
7	Joist splices	21	Piece	2'-0"	1×6	8'	4	6
8	Block bridging	40	Piece	1'-10⅜"	2×6	8'	4	10
9	Closers	12	Piece	10'-0"	1×8	10'	1	12
10	Flooring	800	Board feet	RL	1×6	RL	-	-

inches, and dividing an 8-foot piece of stock into pieces for footers would yield five usable pieces with 11 inches of waste—obviously not economical. To make these calculations more precise (and with less effort), convert the standard lumber sizes and the length in place measurement from a combination of feet and inches into measures of just inches. An 8-foot piece of stock becomes 96 inches, 10 feet becomes 120 inches, and 12 feet converts to 144 inches, while the length in place measurement changes from 1 foot 5 inches to 17 inches. Simple division then shows that dividing a 120-inch length of stock by 17 inches results in seven usable pieces with only 1 inch of waste.

Since the take-off list specifies 45 17-inch (1-foot 5-inch) pieces for footers, ordering seven 10-foot lengths of stock will meet the requirements with enough stock left over to serve as a combination waste, breakage factor, and/or can be put to use on other areas of the project.

Tabular method

Throughout the chapters of this book, equations for determining the necessary materials for different trades will be specifically addressed. Along with the equations for determining materials, many tables will be presented that contain results of these calculations already performed on "sample job sites." These tables can be used as a time-saving device by the contractor in determining sizes, quantities, and totals of materials used for compiling bills of materials and materials lists. For instance, Table 3-2 breaks down by category (per cubic foot of mortar) the material required to mix mortar.

Once the take-off list has been compiled, the next step is to consolidate the information presented on the take-off list into a materials estimate list (Fig. 3-1). By adding together all the pieces of the same size and length of material that appears anywhere on the take-off list (all the 10-foot-long 2 × 6 stock, etc.), a complete total of all the material of that type, size, and length required on the project will exist. Now that a total for each size of each material exists, a waste factor must be determined to be added to those totals.

As a rule of thumb, material of 1 inch or less (flooring, sheathing, etc.), should have a waste allowance of 20 percent

Table 3-2 Quantities and types of material for mortar mixes.

	Mortar mixes		*Quantities*			
Cement sack (1 cu. ft.)	**Hydrated lime or lime putty, cu. ft.**	**Sand cu. ft.**	**Masonry cement, sack**	**Portland cement, sack**	**Hydrated lime or lime putty cu. ft.**	**Sand cu. ft.**
1 masonry cement	-	3	0.33	-	-	0.99
1 portland cement	1	6	-	0.16	0.16	0.97
1 masonry cement plus	-	6	0.16	0.16	-	0.97
1 portland cement	¼	3	-	0.29	0.07	0.86

added to their totals. For materials of 2 inch or better, a 10-percent waste factor added to their totals should suffice. Any quantities of any material needed for the job not specified in the plans should be listed in a separate column on the materials estimate list. For example, if three pieces of 2 × 6 × 8 are needed for temporary bracing purposes, this amount should be indicated in the appropriate column of the list. By adding the take-off quantity, waste allowance, and any non-plan requirements together, a total quantity for each size and length is achieved. The final step in preparing a materials estimate list is to convert this final total into the standard commercial unit of measure. Since in this example lumber is the subject, the previously discussed calculations would be used to arrive at a unit measure of board feet.

The bill of materials

By compiling the information in the stated manner, all components of a bill of materials are now available. In order to present a concise shopping list to one or more material suppliers, the information is now organized as in Fig. 3-2. With each seg-

Item	Size & length	Unit	Takeoff quantity	Waste allowance	Non-plan requirement	Total quantity	Commercial measure board feet
1	6 × 6 × 12	Piece	4	1	–	5	180
2	2 × 6 × 10	Piece	89	9	–	98	980
3	2 × 6 × 8	Piece	15	2	3 (temp. bracing)	20	160
4	1 × 8 × 10	Piece	12	2	–	14	91
5	1 × 6 × 8	Piece	9	2	2 (batterboards)	13	52
6	1 × 6 × RL	Board foot	800	160	–	960	960
7	16d	Pound	–	–	36 nails (framing)	36	–
8	8d	Pound	–	–	23 nails (flooring)	23	–

3-1 *Example of a materials estimate list.*

Item	Quantity	Unit	Size & length	Board measure	Description
1.	5	Piece	6" × 6" × 12"	180	Posts
2.	98	Piece	2" × 6" × 10"	980	Footing/girder /joist
3.	20	Piece	2" × 6" × 8"	160	Spreader/ bridging
4.	14	Piece	1" × 8" × 10"	94	Closers
5.	13	Piece	1" × 6" × 8"	52	Scabs/splices
6.	960	Board feet	1 × 6 × RL	960	Flooring
7.	36	Pound	16d	–	Nails, framing
8.	23	Pound	8d	–	Nails, area coverage

3-2 *One way of organizing a bill of materials.*

ment of the building organized and detailed in such a manner, a bill of material can be furnished to a single supplier or separated and given to any number of suppliers to be filled.

Many different paper forms exist for the use of ordering materials. Figures 3-3 and Fig. 3-4 show only two more types of forms usable for ordering materials from suppliers.

Quantity	Item	Size	Unit cost	Total cost

3-3 *Example of a materials order form.*

Quantity	Thickness	Width	Length	Footage (board ft., linear ft. or square ft.)	Type of material	Cost per/ linear ft. board ft. or square ft.	Total cost

3-4 *Example of a materials list as given to a supplier.*

Specialized computer software

Numerous programs are available to assist estimating, designing, accounting, and managing chores of today's contracting firms. Some software is for sale, and is to be loaded onto your company computer(s) for everyday use. Other programs, found on the World Wide Web, are interactive and provide customized information for a variety of purposes on an "as needed" basis.

This book does not endorse any particular software product over any other. The purpose of this section is to introduce the reader to some of the types of computer software that are available to help in the everyday number crunching of the building trades. Descriptions and recommendations of various software components are presented to assist in the selection of software that will be most useful.

Resident software

Traditionally, software has been a program that is loaded directly onto the hard drive of a computer at the user's location. The user has purchased or licensed the software, or in some cases received it free as "shareware." This type of software can be thought of as resident software since it resides in your computer.

Thousands of different programs exist for just about any need you may have. A simple search of the World Wide Web for "estimating software" yielded 2360 results. Choosing which program(s) to invest in is the hard part. Many of the programs are used on a daily basis and can be linked to other company computers through local area networks (LANs).

Web-based software

Web-based software resides on large, powerful mainframe computers. These computers (and software) are accessible to users via the World Wide Web. Some sites are free of charge while others require nominal "membership fees" before allowing access to their sites.

A Web-based approach has several distinct advantages over the traditional software production and distribution process. First, given the sophistication of Web development tools, the user interface can be designed (and subsequently modified) with considerably less effort, and thus lower cost, than with traditional methods. Second, the cost to distribute the product is minimal. Furthermore, future refinements or additions to the program do not require physical redistribution or reinstallation of the software or documentation. Upgrades and changes only have to be made to the master version (located on the mainframe) for all users to have the benefit of the latest version.

Users with forms-enabled Web browsers (which most are) have access to a seamless interface free of most hardware, software, compatibility, and installation problems. Regardless of the computing resources available at the user's location, each user has access to powerful computational engines residing on the host server.

Administrative software

Contractors have some fundamental needs that require special consideration when evaluating accounting and job cost software from an operations perspective.

It is obvious that building trades are labor-intensive. The need for software that can track and report labor productivity by activity and by crew may not be as obvious. The system should also maintain a database of man-hour productivity and unit productivity by cost code for later use in estimating similar jobs or activities. Most accounting software can report cost by cost code, but not all allow for unit reporting using man-hours and dollars. Still fewer programs allow for maintaining a trend analysis of unit costs.

Another area of interest is work performed on a time and materials (T&M) basis. While most general contractors need to process T&M change orders on their lump-sum contractors, these change orders are often subcontracted out, so there are fewer invoices to track, markup, and bill.

Equipment management

For the trade contractor who has labor, purchased material, equipment, inventory, and possible subcontracted services going to a job, tracking costs and billing them is critically important to ensuring profitability. Software that deals with the contractor's equipment and tools should:

- Track operating and maintenance costs,
- allocate their cost to jobs, and
- schedule maintenance.

These functions call for highly specialized and integrated applications not offered by all software. Purchasing a good general accounting and job cost suite of applications with the hopes of interfacing a stand-alone equipment application is usually not advisable. Integrating additional programs with required applications is so extensive that it would be either marginally successful or very costly, or both.

The equipment management module should track each piece of equipment as a separate record. The software usually performs this function in a manner very similar to the way job costing treats a project. For trade contractors with a significant amount of equipment, the ability to track costs, charge "consumption" to jobs in the form of an hourly rate, schedule maintenance, and track location is very important. In the past, many contractors had to maintain all of this information manually on dry-erase boards, index cards, or by using custom-made spreadsheets. With the advent of Equipment Application software, all of these functions can be handled efficiently from one centralized program.

Service management

For mechanical and electrical contractors, it is common to have a service department that makes service calls to maintain or repair equipment installed by the contracting division. This significant aspect of a trade contractor's business requires special consideration. As with Equipment Management, it is not advisable to purchase a separate Service Management program and attempt to interface it with other applications. Because the software needs to work closely with and handle processes like call management, dispatching, billing, and tracking service history, integration is primary.

Trade contractors are likely to be more departmentalized than the typical general contractor, diversifying into multiple types of work and geographic areas. Much of this activity requires accumulated financial information, which needs to be reported separately. In many cases, one job will involve more than one department. Because of this interdepartmental involvement, the ability for the system to maintain separate business units and handle interdepartmental charging is important.

Designed specifically for service departments with numerous trucks, inventory, service calls, maintenance contracts, and work orders, there are other appropriate uses for service management applications. Anyone who does on-site maintenance and has many small jobs, especially industrial contractors, could find this application effective and efficient.

Before the service management application became available, service calls were either set up within the Job Cost application or tracked within a separate department in the General Ledger. The latter option provided departmental profitability, but no discrete tracking of cost or profit by service call. In addition, with the former alternative, setting up a job record for each service call usually proved to be too time-consuming.

Ideally, service management software should begin tracking the call as soon as it comes in: automatically converting the call to a work order when dispatched, and converting it again to an invoice and a service history record when the work is completed. Tremendous gains in efficiency and a reduced risk of missed costs result when this integrated, streamlined process is used.

Labor productivity

With labor being one of the most significant and variable costs for a contractor, an adequate system for tracking productivity and cost is crucial. This may sound basic, but you will find that different job cost applications maintain and report labor productivity differently. Fundamentally, a contractor needs to track:

- budgeted, actual, and forecasted costs,
- man-hours, and
- units for all labor cost codes (activities).

With this basic data, the system should be able to provide man-hours per unit and cost per unit either as a report or through inquiries. It is an added benefit if the system can track trends over several weeks and/or compare unit costs for a given cost code using several similar jobs.

Human resources

Another new application for some contractors is Human Resources (HR). For some time, HR was nothing more than a few fields and screens added to the Payroll program. However, HR requirements have become so complex that contractors now need to track individual applicants, safety records, vacation time, benefits, salaries, and performance review history. For that reason, software exists with a separate application for just those functions. Need for this application is dependent on the number of employees, the amount of hiring and layoff activity, the complexity of benefits, and the general sophistication of the HR function within your company. Look for complete integration with the Payroll application so that duplicate maintenance of employee master records is not required.

Time & material contracts and billing

Trade contractors normally perform some if not all of their work on a time and materials (T&M) basis. Consequently, the contractor needs a system that maintains costs for a job using the standard cost-code structure used by the contractor. However, the system may also need to track costs according to a different and separate billing structure. It needs the ability to maintain billable costs and non-billable costs separately for billing purposes, but collectively for costing purposes.

The ideal software needs to maintain accurate billing of T&M jobs and accurate cost tracking by cost code. Most T&M billing functions are a subset of the billing application and receive data directly from job cost.

General ledger

A few General Ledger functions are unique and deserve an individual mention. One, in particular, is the ability to process transactions for multiple companies through one system. This is usually an issue for larger contractors with multiple operating companies that regularly transact business with each other. If this is the case, you will need software that allows you to designate the company (as well as the division and account number for journal entries) and to have the system automatically create the appropriate intercompany transactions. The system should also allow for separate and consolidated financial reporting as needed.

Multicompany capability needs to be distinguished from multidivisional accounting within one company. A much simpler task, most systems allow transactions to be posted to a single division or split among multiple divisions. Any trade contractor with more than a single operating division should use this important feature. With a reasonably sophisticated General Ledger application, management should be able to regularly get income statements to the contribution margin level (gross profit less allocated overhead) for each division, whether they are by product/service, geography, or both. This requires some thought during the initial setup, but then very little extra effort on a month-to-month basis.

Change order management

A system's ability to manage change orders throughout the construction process is an important issue. The process usually includes:

- identification,
- estimating,
- submittal,
- cost tracking, and
- billing.

Some software includes the Change Management application within the Job Cost program. Other programs are offered as separate modules or stand-alone programs. The sophistication level varies considerably from program to program; be sure to compare the functionality of each program carefully. Consider your particular needs and weigh them against the features of the program.

A feature that will always be beneficial is the ability to interact with other programs, exchanging information between applications. For example, save man-hours on repetitive tasks like transferring simple information (customer name, job site location, telephone contacts, customer number, and job number) from one application to the other.

Purchasing

Early software provided no option for tracking contracts with service companies. Contractors were forced to use the Purchasing application to issue and track these contracts. With the development of more sophisticated Subcontract Management applications this is no longer the case. However, for large quantity material ordering or equipment rental for projects, the Purchasing application remains the correct program.

One of the important functions of the purchasing software is posting the committed cost to the Job Cost application. This keeps project managers aware of all commitments against their jobs (without requiring that invoices be received and posted). The benefit in this circumstance is a quicker response to cost overruns, limiting unnecessary expense.

The processing of purchase orders through an integrated system provides another important, but less obvious, benefit. Most Purchasing systems post the purchase order (PO) detail into the invoice entry screen when the PO number is entered. This saves the accounts payable clerks from entering the vendor, line item detail, job, and cost codes.

Task specific software

With new software coming on the scene almost daily, it is not hard to find a computer program for almost any application. Estimating and designing software is in abundance. The brief descriptions of various software given here are to provide you with a taste of what is available.

As mentioned in Chapter 1, personal computers are now able to perform a variety of functions within the realm of the construction industry. Some of the operations a computer performs best are numbers crunching, memory storage, and information organization, which are precisely the elements involved in putting together a bill of materials.

One of the specialized software programs available today allows the contractor to enter information in the form of measurements, location, and design type, among other details used by the program to generate an estimated bill of materials. This bill of materials can be generated in a variety of different final forms, but one of the more common and popular forms is based on the Construction Specifications Institute's (CSI) Masterformat. The CSI Masterformat is comprised of sixteen divisions:

Division 0:	Bidding Requirements
Division 1:	General Requirements
Division 2:	Site Work
Division 3:	Concrete
Division 4:	Masonry
Division 5:	Metals
Division 6:	Wood and Plastics
Division 7:	Thermal and Moisture Protection
Division 8:	Doors and Windows
Division 9:	Finishes
Division 10:	Specialties
Division 11:	Electrical
Division 12:	Furnishings
Division 13:	Special Construction
Division 14:	Conveying Systems
Division 15:	Mechanical
Division 16:	Electrical

A program operating on the basis of the CSI Masterformat is particularly beneficial to both the general contractor and each individual sub-contractor. Because each individual trade can utilize this program by plugging in data specific to his own trade, the results will be very accurate for each trade. Meanwhile, the general contractor, or architect can utilize the system to gain an overview of the project utilizing information which is, for all practical purposes, very accurate.

The software is so sophisticated today that many of the variables within each of the sixteen divisions can be changed in order for the contractor to perform "what-if" analysis. That is, after entering all of the necessary data, one or more of the variables may be changed in order to observe how the changes affect the project in whole or in part. This process can be repeated as often as the user of the program wishes.

Another benefit of this type of software is that, with the appropriate software, CSI division 2 (sitework) is reported as a separate item in order for a contractor to focus on issues concerning the building without being affected by the sitework. On the other hand, a contractor who is primarily concerned with the site work can concentrate on just the division 2 results without considering any of the other divisions.

Normally, as part of product support services, updated information is periodically sent to the software user from the manufacturer in order to adjust the program's data files. These data files contain information such as the current rate of inflation in different areas of the country and regional differences in labor and materials costs. Using the latest information available for these adjustments provides the contractor with timely costs for materials, labor and even borrowing money.

HVAC

In the building industry, particular emphasis is placed on the energy use of the HVAC systems of a building. The purpose of building energy analysis (BEA) is to compare the energy use and operating costs of alternate system designs in order to choose the optimal design. The analysis mathematically simulates the thermal performance of the building to determine cooling and heating loads. It then mathematically simulates the performance of HVAC equipment in response to these loads to determine energy use over the course of a year. Finally, energy data is used to calculate operating costs.

A wide variety of building energy analysis methods are currently available to builders, contractors, and HVAC engineers. These methods range from the simple to the sophisticated. The simplest methods involve the largest number of simplifying assumptions and therefore tend to be the least accurate. The most sophisticated methods involve the fewest assumptions and thus can provide the most accurate results. Generally, BEA methods are divided into three categories:

- Single Measure Methods (example: Equivalent Full Load Hours)

- Simplified Multiple Measure Methods (example: Bin Method)
- Detailed Multiple Measure Method (example: Hour-by-Hour)

Single measure methods

These methods involve one calculation of annual or seasonal energy use. For example, the degree-day method calculates energy use by combining one degree-day weather value with a load value and an efficiency value to obtain seasonal or annual energy use.

Similarly, the equivalent full load hour method combines full load capacity, full load efficiency, and equivalent full load hours to obtain annual energy use.

In both cases, this level of simplicity is achieved by using such sweeping assumptions that the accuracy and reliability of these methods are very limited.

Simplified multiple measure methods

These methods involve calculations of energy use at several different conditions, for example, the Bin Method, where energy use is computed at a series of outdoor air dry-bulb conditions. Results are then weighted according to the number of hours each dry-bulb condition is expected to occur to determine annual energy use.

Example: 47 degrees Fahrenheit would be used to represent the range of conditions between 45 degrees Fahrenheit and 50 degrees Fahrenheit and is referred to as a "bin." Building loads and equipment energy use are first calculated for the 47 degree Fahrenheit bin. To determine annual energy use for that bin, energy results are then multiplied by the number of hours per year that the temperature is expected to be between 45 and 50 degrees Fahrenheit. Similar calculations would then be repeated for all other temperature bins for the local climate and would be summed to determine overall annual energy use.

While the Bin Method provides a vast improvement in sophistication over single measure methods, it has a fatal flaw. This flaw is that it must separate weather conditions, loads, and system operation from time.

Example: The number of hours in the 47 degree Fahrenheit bin (when the outdoor dry-bulb is between 45 and 50 degrees Fahrenheit) occur at various times. These temperatures are reached at different times of the day and night, as well as on different days of the week, weeks of the month, and months of the year. Because a single calculation is performed to represent energy use for all these different times, it is almost impossible to accurately:

- Link solar radiation and humidity conditions to the bin,
- Consider hourly and daily variations in internal loads,
- Consider the transient hour-to-hour and day-to-day thermal performance of the building, and
- Predict time-of-day energy use and peak demands.

Inevitably averaging assumptions must be made to fit all these considerations into the framework of the bin analysis. This process does impair accuracy to varying degrees.

Detailed multiple measure methods

These methods perform energy calculations on an hour-by-hour basis. Consequently, they have the potential to satisfy all the requirements for high quality energy analysis results. There is, however, a certain amount of variation among different detailed multiple measure methods, leading some methods to meet the accuracy requirements better than others do. Within the detailed multiple measure category are two major subcategories: *the Reduced Hour-by-Hour Method* and *the 8760 Hour-by-Hour Method.*

Reduced hour-by-hour method

This method typically uses one 24-hour profile of average weather conditions per month. Energy simulations are performed for this average profile and results are then multiplied by the number of days in the month to obtain monthly energy totals. Upon this foundation, different reduced hour-by-hour methods make various improvements to enhance the accuracy of results.

Some methods analyze building operation for a typical weekday, Saturday, and Sunday each month since building use profiles differ significantly between these days. One average weather profile is still used for all three typical day simulations each month. Some methods also analyze equipment operation for a hot and cold day each month in an attempt to improve estimates of peak electrical demand. Some methods simulate building operation for one 7-day week each month to try to account for day-to-day building dynamics. However, one average weather profile is still used for all seven days of the simulation.

The fundamental principle of this method is averaging. This is assuming that building and equipment performance on hotter and colder than normal days each month averages out, so that monthly energy use can be accurately predicted by simulating a small group of days using average weather conditions.

8760 hour-by-hour method

This method simulates building and equipment performance for all 8760 hours in the year using the proper sequence of days and actual weather data. No weighting of results or use of simplifications is necessary. The fundamental principle is to produce the most accurate energy and operating cost estimates by simulating the real-time operating experience of a building over the course of a year. All the requirements for high quality energy analysis results can be met with this approach.

The actual weather data accounts for the range and timing of weather conditions in detail. Further, the hourly and daily variation of building occupancy, lighting, and equipment use is easily accounted for. In addition, the full year simulation tracks the dynamic hour-to-hour and day-to-day response of HVAC equipment to the thermal behavior of the building. The ultimate result is high-quality data that can be utilized to produce accurate, detailed data about the quantity and timing of energy use. Both are requirements for accurate operating cost estimates.

Duct sizing

Software for designing duct size is available. Examples include menu-driven calculation utilities designed to simplify the process of sizing and specifying ductwork for a variety of HVAC applications. Static regain, equal friction, or constant velocity methods can be used to determine the optimal duct sizes for round, rectangular, or flat oval ducts. Optionally, the duct height and width constraints to control sizes can also be specified. This can be used for analyzing problems in existing systems where the duct sizes are already specified.

Noise levels and required attenuation are also printed for each runout duct. A library of fan data used for noise calculations is built into the program. Programs offer a variety of comprehensive and concise printed reports and bills of materials containing both labor and material costs. Up to 500 duct sections are allowed, and the program is suitable for both constant air volume (CAV) and variable air volume (VAV) systems. Data is generally entered using a series of simple, fill-in-the-blank screens. In some programs, the input data can be taken directly from a duct drawing file created in AutoCAD.

Computer aided design

NOTE: Computer aided design (CAD or AutoCAD) is becoming a more frequent resident on contractor's computers. This design system allows much freedom, accuracy, and convenience. Applications software that is compatible with CAD programs performs additional functions and should receive higher consideration than software that is not.

Modeling techniques for building services

Programs compatible with CAD systems can perform various modeling functions to assist in selection and application decisions. Characteristics of these programs should include:

- Works inside AutoCAD
- Uses BS 1192/AEC layering conventions
- Symbols are parameter-driven
- Easy menu selection from toolbars
- 2D/3D switch for drawing and visualization

The Services Environment will yield productivity gains. It offers unparalleled scope for modeling building services design. The combination of CAD with CAD compatible specialty programs create a comprehensive system that takes you from the initial drawing stage through to complete design, working drawings, and 3D visualizations.

The power of these combinations allow the user to:

- Calculate cooling loads, place grilles to supply the air quantity worked out by the software and then route the ductwork through the ceiling void back to the fan. Calculated duct sizes can then be automatically transferred onto either a 2 line drawing or 3D model.
- The process can be repeated when any design parameters are altered, ensuring that a direct relationship between the design and drawing is maintained. A similar approach is taken for pipe systems, lighting and cabling.
- User-controllable schedules of radiators, pipe work, valves, ductwork, luminaires, or wiring are taken directly from the calculated results and can be placed anywhere on the drawing.
- Plant items such as sectional boilers, modular boilers, flues, headers and calorifiers should be produced parametrically to enable exact dimensions to be selected for individual manufacturers.
- Items that form part of the pipe system such as tees, bends, pumps, controls, valves, flanges, and sanitary fittings should be automatically provided.

A wide range of air-handling equipment is normally offered such as:

- Filters
- Heating and cooling coils
- Fans, grilles, and dampers
- Duct fittings—radius bends, square bends, branch shoes.

Where ductwork is entered manually, the dimensions of elements should be based on previous items (to minimize data entry time). Conduit, trunking, cable tray, ladder rack and control panels sized to manufacturers dimensions are standard.

Light fittings and a range of electrical symbols including fire alarms, sockets, switches, bells, break glass units and many more are handled in detail. Luminaire references are offered on insertion. For schematic diagrams the program should offer a range of circuitry symbols.

Estimating labor costs

Properly estimating labor costs involves various components that must be considered. It is important that these components and their terms be clearly understood before any estimating takes place.

Direct labor cost

Direct labor costs are normally calculated using the formula: Labor Cost equals Quantity divided by Production Rate multiplied by Labor Rate, or

Labor Cost = (Quantity/Production Rate) Labor Rate

Quantity: The amount of work (in measured units) that needs to be done. For example, when framing a house the unit of measure would be number of square feet that needs to be framed.

Production Rates: Production rates or Labor Productivity represents the units of work produced by a person or crew during a specified period of time.

Labor Rates: Labor rates for this application refer to the hourly wages paid to an individual.

Both Production Rate and Labor Rate information are generally obtained from one or more of the following sources.

- Personal Experience/Company Records
- Commercial Sources
  1. Richardson's Estimating Guide
  2. The National Construction Estimator
  3. Means Building Construction Cost Data
  4. Walkers Estimating Handbook
- Scientific (computer) Modeling

Note, however, that labor rates set forth in the National Construction Estimator include all benefits. Accordingly, the estimate would have to add employment taxes, health insurance, and the other identified applicable benefits so that the wage expense estimated is computed on the same basis.

Productivity factors

Productivity factors will affect the production rate in the calculations. These factors are real-world influences that, in part, determine the amount of work an individual or a team can produce during a measured period of time. These factors are

- Variation in the skill level of the laborers,
- Variation in project complexity from site to site,
- Climate—how heat, cold, humidity affects laborers, and
- Quality of supervision.

Indirect labor costs

Indirect labor costs are all the taxes, insurance, and benefits that both employee and employer end-up having removed from their pockets. Most common among these items are

- Social Security Tax,
- Unemployment Tax,
- Workers Compensation Insurance,
- Public Liability and Property Damage Insurance, and
- Fringe Benefits.

Table 3-3 provides a labor cost example including direct and indirect labor costs and how they affect both employee and employer.

Worker hours required to frame 1000 square feet of living space: This is the total hours required by a skilled carpenter to frame a basic structure of this size. Hourly labor cost: This includes the basic wage, employer's contribution to welfare, pension, vacation, apprentice funds, and all tax and insurance charges for a skilled carpenter.

Table 3-3 Labor cost example.

Description	Employee	Employer
Wage rate	$20.00 per hour	$20.00 per hour
Federal tax (15%)	−$3.00 per hour	—
State tax (3%)	−$0.60 per hour	—
Social security (7.5%)	−$1.50 per hour	$1.50 per hour
Unemployment (3%)	—	$0.60 per hour
Workers compensation (15%)	—	$3.00 per hour
Fringe benefits (10%)	—	$2.00 per hour
Public liability insurance (2%)	—	$0.40 per hour
TOTAL	$14.90 per hour	$27.50 per hour

Worker hours

For the purposes of example, consider the worker hours required for framing a typical framed structure. Table 3-4 itemizes the components included in a typical framed structure along with the worker hours necessary to measure, cut, and install the wood for that component in the structure. These estimates should remain constant from year to year unless new framing techniques are developed. As shown in Table 3-4, it takes a skilled carpenter about 221 worker hours to frame a 1000 square foot (SF) building.

Table 3-4 Worker hours per 1000 SF of floor space.

Sills, per blocks, floor beams	16
Floor joists, blocking, bridging	26
Subfloor, plywood	10
Studs, plates, headers, blocks, bracing	80
Wall sheathing, plywood	13
Ceiling joists, beams, trimmers, blocks	36
Rafters, collar beams, ridge boards	26
Roof sheathing, plywood	14
Total hours to frame 1000 SF of floor space	221(est.)

To compute the average hourly labor costs per square feet (SF) of framing, the preceding information is used as follows:

Work hours to frame 1000 square feet of living area (est.)	221
Carpenter's labor rate per hour (est.)	× $29
Labor cost to frame 1,000 SF of living area (est.)	$6409
Divided by number of SF framed	1000
Cost of labor per SF of framing (est.)	$ 6.41

Adjusting the numbers for regional and updated information, the answer arrived at represents an industry average of the carpenter's labor cost to construct 1 square foot of residential living space.

Multiply the hourly rate by the number of square feet of living space framed for either the entire year or for a specific job. This will give you an estimate of the labor cost that should have been incurred for the project(s).

To properly estimate labor costs on any construction job, there are two components that must be considered:

- Man-hours required to perform the work.
- Taxes and insurance costs associated with the total man-hours allocated to the job.

The first consideration when estimating labor costs is to determine an estimate for the number of man-hours required to perform an actual job. In order to demonstrate how complete labor cost estimating should be, the example of a drywall installation will be used here: however, the same procedure, thoroughness and formulas can and should be used for calculating labor costs in any of the construction trades.

All segments of the installation should be estimated separately. Ceilings, walls, and finish work are the easiest segments of the job to remember to include in estimates, but any and all work required to correct any framing imperfections must be included as well. Before estimating labor for the installation of the drywall panels, carefully inspect the framing and make allowances for any labor costs necessary to correct framing problems. Correcting problems to ensure that the framing is plumb and level can require as many (if not even more) man-hours as the installation of the drywall panels themselves.

To calculate the labor costs of installing gypsum drywall panels, use this simple formula: the *labor rate per hour* multiplied by the *total number of man-hours* equals the *total labor cost.*

At this point, the definition of the term *man-hours* should be clearly understood. For example, if two men require 10 minutes to install a 4 × 8 ceiling panel, including fasteners but not including taping and finishing; and the ceiling requires three panels, the total labor time equals 1 man-hour (10 minutes multiplied by three panels = 30 minutes; then, multiplied by two men = 60 minutes = *1 man-hour*). When estimating labor costs, always remember to multiply the number of hours required to perform the job by the number of people performing the job to obtain the actual number of "man-hours" it will cost to do the job.

Once the total number of man-hours has been calculated, multiplying that figure by the labor rate per hour will result in the labor cost estimate to complete the basic installation for that segment of the job. With this formula, and the materials estimate, calculations can be completed for each individual segment of the job.

Walls

Once estimates have been completed for ceiling areas, the same estimating procedure should be used for the wall segments of the job. Unless otherwise specified by the general contractor, architect, or owner of the property, the drywall contractor has the option of using either nails or screws for the installation of gypsum board panels over wood framing. If screws are to be used, less labor will be required since fewer fasteners are needed per panel; however, screws cost more than nails, so a portion of the savings in labor costs is absorbed by higher material costs. Still, although screws have a higher initial cost than nails, they also have much greater holding power; consequently, by absorbing a higher materials cost up front, the contractor may very well reduce her overall expenses resulting from repairing loose panels caused by nail pops and fastener failure during construction or in the future. The expense of repairing loose panels due to fastener failure is, of course, a variable cost that the drywall contractor must calculate using past experience as a guide.

If nails are used, the initial materials cost will be less, but labor costs will be higher because more nails than screws are required per panel. The variable cost of repairing loose panels due to fastener failure must again be considered. (Of course, if gypsum board panels are being applied over metal framing, no choice of fasteners exists; screws must be used.)

Cut outs & corners

Another consideration in estimating labor costs for wall applications is the number and size of window, door, electrical outlet, and switch openings that must be cut into the gypsum board. Each of these openings must be cut precisely and requires additional time to ensure a quality job. Be sure that an accurate count is taken of the number and type of openings that need to be cut, and add these costs to the total labor estimate.

The installation of corner bead is another component of the drywalling job which adds additional labor time and expense. A careful study of the job specifications needs to be made to calculate the appropriate linear footage of corner bead to be installed.

Gypsum drywall finishing

Taping and filling the joints, spotting the fastener heads, and applying texturing or other finishing material to the gypsum boards are all components that fall into the category of finishing labor. The drywall contractor's familiarity with her crews and accurate records of the amount of time each crew required to perform these tasks on previous installations will enable the drywall contractor to properly estimate the number of manhours needed for this step of the job.

Scheduling & avoiding downtime

Whenever joint compound is applied during the application of gypsum drywall panels, work on that section of the job must be stopped until the joint compound is completely dry. This work stoppage, known as *downtime*, can result in additional expenses to the drywall contractor if he has to pull crews off the site and move to another job before returning to finish the first job. To avoid this situation, plan to have different rooms at different stages of completion throughout the installation process. A crew should be able to embed reinforcing tape, spot fastener heads, or apply texturing in one room, and while waiting for the joint compound or texturing to dry, move into other rooms to fasten panels onto framing and ceilings or apply successive coats of joint compound or texturing. Once the panels are up and the joints are taped, fastener heads spotted, etc., the crews should be able to move back to the first room(s) to apply any necessary successive coatings of joint compound or texturing. No matter how well the work is sequenced to minimize downtime, some period of downtime will most likely be experienced. Make sure to inform the customer of the reason for this downtime, so that he will not fall under the misconception that he is being neglected simply because the drywall crew(s) is not on site.

Taxes & insurance costs

Labor costs are actually composed of two sets of costs. In addition to the manhour costs already discussed, another factor referred to as the *labor burden* must now be added in order to calculate total labor costs. On average, labor burden will increase labor costs by 25 to 30 percent on each construction job. This labor burden is the percentage of payroll dollars the contractor is compelled to pay to government agencies in the form of taxes and in the form of premiums paid to insurance companies. The following taxes and insurance are examples of the components making up the labor burden:

- Unemployment insurance. All states levy an unemployment insurance tax on employers, based on total payroll for each calendar quarter. The actual amount assessed will vary on the history of unemployment claims filed by employees of the company. This tax may range from 1 to 4 percent of payroll dollars.
- FUTA, the Federal Government's unemployment tax. This tax has averaged 0.8 percent of payroll dollars per

employee up to a maximum dollar amount per year periodically set by law.

- Social Security and Medicare (FICA). The Federal Government also requires FICA payments averaging 7½ percent of payroll dollars per employee up to a maximum dollar amount per year prescribed by law.
- Workers' Compensation insurance. All states require this insurance be carried by employers to cover employees in the event of a job-related injury. The cost of this insurance is a percentage of payroll dollars based on the occupation of the employee. Secretarial jobs have a low rate compared to higher-risk jobs such as construction work, which carries a rate usually between 5 and 10 percent of payroll dollars.
- Liability insurance. This insurance protects the contractor in the event of an accident. Most general contractors will require that a certificate of insurance be presented by subcontractors prior to beginning work. The cost of this insurance is based on total payroll and is dependent on the location of the company (and the work), history of claims by the company, and the liability limits needed.

From these examples, it is clear why labor burden must be included in every labor cost estimate. Because each State and Local Government passes their own laws and sets their own percentages for these mentioned taxes and individual companies regulate insurance rates, frequent updating of information pertaining to these taxes and insurance costs must be made to avoid under-pricing jobs, reducing profit margins, and perhaps even losing money.

Review questions

1. If two full-time employees accomplish 20 tasks in a week, how many such tasks will 5 employees accomplish in a week?
2. Two houses were framed, with each home consisting of 2000 SF. The estimated cost of labor (framing only) is $6.41 per SF. What is the estimated cost of labor (framing only) for each house? For both houses?

3. A 3500 square foot house is framed with a total cost (framing only) of $20,930. What is the per square foot labor cost?
4. Name three things that software dealing with the contractor's equipment and tools should provide.
5. A two-man framing team averages 15 square feet per hour, (just framing). Their next job totals 2000 square feet; how many man-hours will the total job take?

4

Grading and excavating math

The term *grading* generally refers to the process of relocating, removing from, adding to, and reshaping the surface of a construction site. This process is undertaken in order to obtain the desired elevations of the surface at, and adjacent to, a location on the site where a structure will be erected. Once this has been accomplished, further earth moving is usually required. If, for example, the structure is to have a sub-grade (basement) level, the earth lying within the building lines must be removed down to the floor level, as specified in the plans, less the thickness of the floor paving and subfill. Once the earth is removed, further earth may have to be removed for footings under the foundation walls. All of this rearranging and reshaping of earth comes under the general category of excavation.

Determining grade

The plot plan of a building site will normally include gradient lines that establish the elevation of different areas of the site. These contour lines (Fig. 4-1) connect points of equal elevation and are marked with the elevation of that point relevant to a reference point. Generally, the plot plan will be drawn by the architect or draftsperson, who will establish and mark all relevant gradients; however, that does not mean that the contractor should not know the ins and outs of establishing, measuring, marking, and working with elevations.

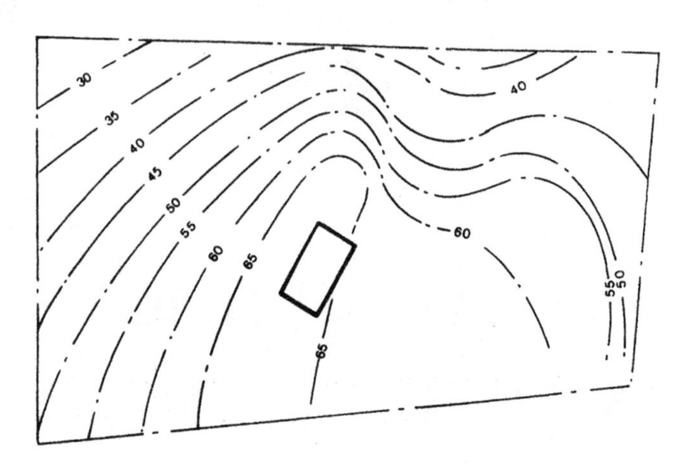

4-1 Contour lines depicting various land elevations.

In order to reshape the site to the grade(s) specified in the plans, a contractor will begin the excavation phase of a job by surveying the site. Using a transit, he/she will make many readings and calculations enabling him/her to stake out the site in such a manner so as to appropriately guide the earth movers.

By knowing the distances from the location of the transit to various objects, the contractor can take readings through the transit and, with the help of some trigonometry, derive the relative elevations of high and low spots throughout a construction site.

Measuring elevations

To begin with, an object's relative elevation is the vertical distance above or below a reference plane (an established height, very often taken as sea level) that has an actual or an assigned elevation (sea level has an assumed elevation of 000.0 feet). On construction sites, this reference point is usually some permanent structure such as a tree, manhole cover, or the surface of a finished road and is referred to as a *bench mark*. The differential method of leveling entails measuring the difference between the elevation of the bench mark and the elevation of some other point. Once the measurement has been taken, the contractor can compare it to the bench mark elevation to determine the elevation of the point in question.

Measuring grade

Although the term *grade* can have several meanings, it is used here to indicate the elevation of surfaces at different points of reference. In preparing the actual building site, creating a level area corresponding to elevations laid out in the plot plan will probably be the first order of business. Removing earth from particular areas (cut) to bring them down to the required level or adding earth to other areas (fill) to bring them up to required levels are fundamental operations of excavation.

In order to ensure that the earth-movers cut and fill in the appropriate areas and in the correct amounts, numerous grade stakes are used to indicate to the equipment operators whether to cut or fill and how much. A shorthand language of sorts is used on these stakes, not only for brevity but for clarity. Machine operators need to be able to clearly identify the information on each stake in order to properly grade the area. For example, a stake with "GRD" written on it means that the surface is currently at the required level and no cutting or filling is necessary.

An area requiring cutting or filling will have a stake with either a C or an F (C = "cut" and F = "fill") on it, along with a numerical value indicating the difference between the required elevation and the existing elevation. For clarity sake, these numbers are written as a whole number with any fraction expressed as a decimal which is raised and underlined. For example, if an area has a surface 3½ feet higher than the necessary elevation, the corresponding stake will read "C $3^{\underline{50}}$"; an area requiring fill of 3½ feet will have a stake reading "F $3^{\underline{50}}$."

The plot plan will dictate the necessary grade elevations throughout the building site, in order to properly mark and set out each stake the existing elevation of the surface must be known. Again differential leveling effectively determines existing surface elevations; however, for this purpose, a number of settings are possible from a single instrument location. Calculations can be speeded up in this instance by using grade rod and ground rod values.

When using a transit, the height of the instrument is always to be taken into consideration. The grade rod represents the difference in elevation of the required elevation and the height of the instrument; while ground rod is the rod reading of a particular point as seen in Fig. 4-2. The following are rules of thumb to keep in mind when sighting elevations:

1. On level surfaces of equal elevation, grade rod measurements will be the same across all points surveyed from the same instrument location.
2. The instrument height will always be greater than the elevation of the existing surface at any point on the site, however, it may be less than (Fig. 4-3) or greater than (Fig. 4-4) the required plot plan elevation.
3. If the instrument height is less than the elevation required by the plot plan for that area, the difference will equal the sum of the grade rod and ground rod and the area will need to be filled.

   For example, in Fig. 4-3, the specified grade level is 150.50 and the height of the instrument is 131.0; therefore the grade rod is 19.50 (150.50 − 131.0 = 19.50). The ground rod is 5.5 (the rod reading of the point on the rod intersected by the plane of the transit). By adding the ground rod to the grade rod (5.5 + 19.5 = 25) it can be seen that a stake marked "F 25^{00}" should be placed at this location indicating the correct amount of fill needed.
4. If the instrument height is greater than the required plot plan elevation called for, the difference between the ground rod and the grade rod will equal the amount of cut or fill required.

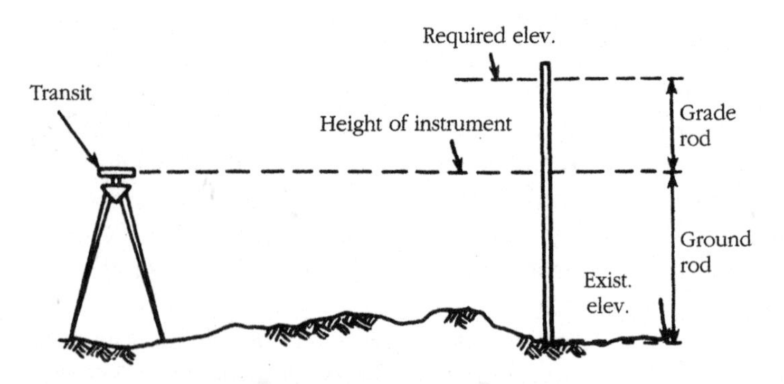

4-2 *Grade rod and ground rod values.*

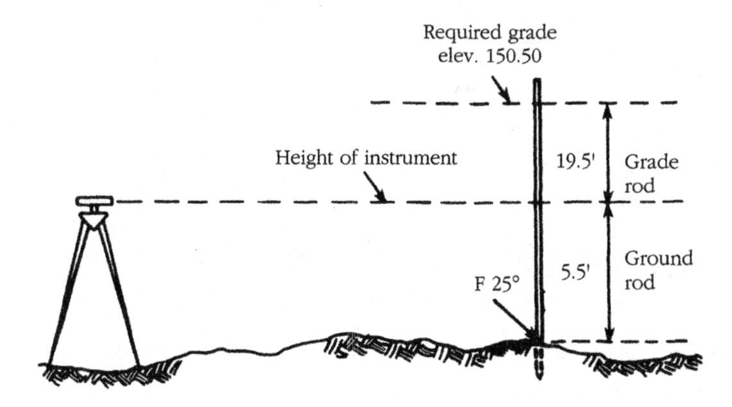

4-3 Elevation of grade less than plot plan elevation.

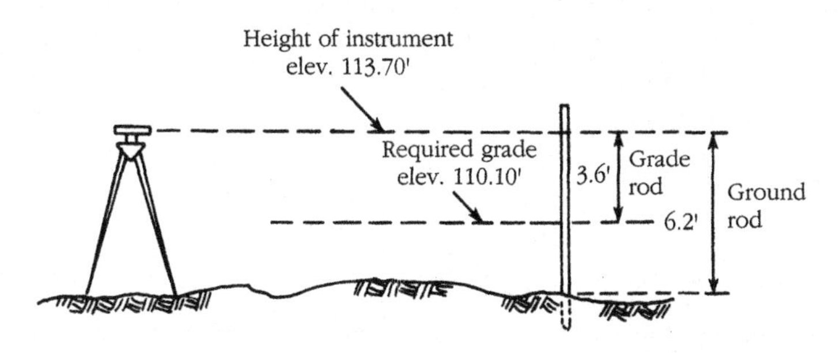

4-4 Elevation of grade greater than plot plan elevation.

As can be seen in Fig. 4-4, where the ground rod measurement is larger than that of the grade rod, fill would be needed to increase the existing surface elevation to a level meeting the required plot elevation. The amount of change needed would be "F $2^{\underline{60}}$" (6.2 − 3.60 = 2.60).

If, as in Fig. 4-5, the situation is reversed by having the grade rod measurement larger than the ground rod measurement a cut would be indicated. The amount of the cut would again be determined by the difference between ground rod and grade rod. As seen in this example the measurements and the resulting equation would be 8.20 − 5.40 = 2.8 or a cut of 2.8 (C $2^{\underline{80}}$).

Estimating the number of cubic yards to be removed from a surveyed parcel of land uses several simple calculations.

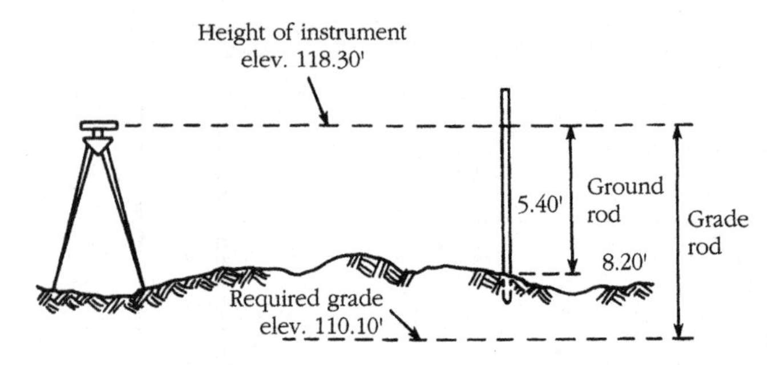

4-5 Elevation differential requiring surface level to be reduced (cut).

First, estimate the average height of the station elevations from the given bench mark. Although averaging is a statistical function, it is very easily done. Begin by subtracting the height of the bench mark from the station elevations: For example, referring to Fig. 4-6:

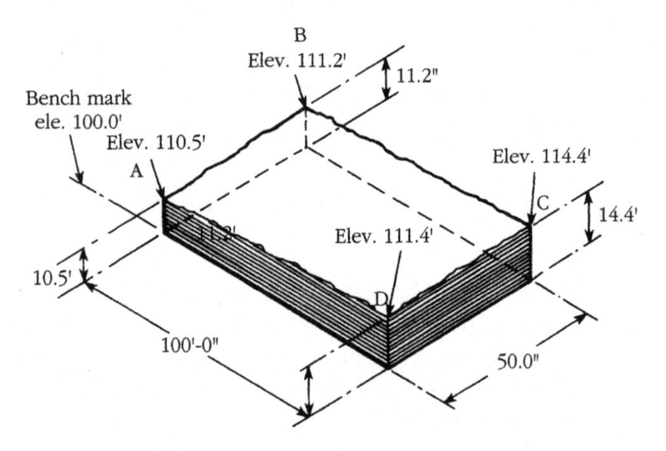

4-6 An example of bench mark and station elevations.

1. Station (a) has an elevation of 110.5; subtracting the height of the bench mark (100) leaves a station elevation of 10.5.
2. Station (b) has an elevation of 111.2; subtracting the height of the bench mark (100) leaves a station elevation of 11.2.

3. Station (c) has an elevation of 114.4; subtracting the height of the bench mark (100) leaves a station elevation of 14.4.
4. Station (d) has an elevation of 111.4; subtracting the height of the bench mark (100) leaves a station elevation of 11.4.

Adding the station elevations together as below:

$$10.5 + 11.2 + 14.4 + 11.4 = 47.50$$

Next, divide the sum of the station elevations by the number of stations (a,b,c,d = 4 stations):

$$\frac{47.50}{4} = 11.875 \text{ feet}$$

The average elevation of the stations is 11.875 feet. Next, multiply the square footage of the parcel, (the length of the parcel multiplied by the width of the parcel) by the average height of the stations, which in this case would be:

$$5000 \times 11.875 = 59{,}375 \text{ cubic feet}$$

The standard terminology of excavating refers to measures of cubic yards, so there is one more calculation to perform. Converting 59,375 cubic feet into cubic yards. Since a cubic yard equals 3 feet of height multiplied by 3 feet of length multiplied by 3 feet of length; one cubic yard equals 27 feet, consequently the conversion is:

$$\frac{59{,}375}{27} = 2{,}199.07 \text{ cubic yards}$$

Plotting areas

Staking out a rectangle that corresponds to the exterior of the structure to be built is the next procedure to be done, following the determination of the location and alignment of the building on the site. Using a transit the procedure includes:

1. Define an established reference line to represent the lateral limit for one of the front corners of the building. This can be a property line, street line, etc., as represented in Fig. 4-7 by line AB.

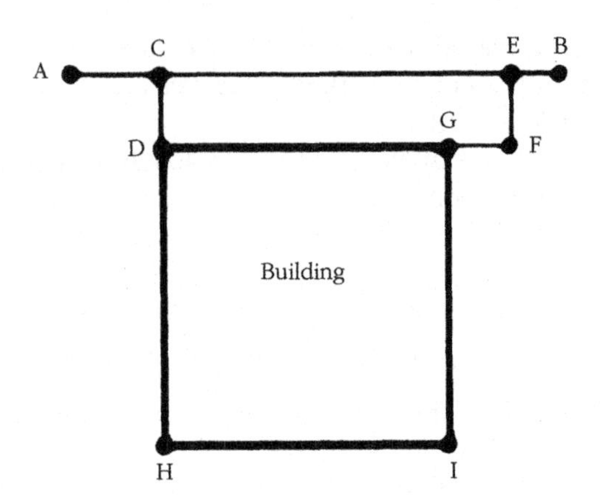

4-7 *The establishment of the lateral limits of the front corners using a reference line.*

2. Locate the transit at point C and determine point D as a front corner.
3. Move the transit to point E along line AB (a distance greater than the intended length of the project) and determine point F at a distance from AB equal to the distance of D from AB.
4. Beginning at point D, measure out the front of the project along line DF, establishing a front corner G opposite front corner D
5. Reposition the transit to point C and shoot E. Then, swiveling the head of the transit 90 degrees (toward what will be the rear of the building) determine point H, a rear corner of the building.
6. Locate the transit at point G and shoot D. Then swivel the head 90 degrees (again toward what will be the rear of the building) and establish point I, the remaining rear corner of the building.
7. Relocate the transit to point I and sight H. If IH is identical to GD the work has been proved and is correct.

More and more often these days, structures do not directly correspond to a simple rectangle. Structures comprised of irregular shapes require a bit more care when laying out their outlines. A rectangle representing the major portion of the building is laid out first, using the same techniques as earlier to establish each corner. Any irregularities are then plotted as smaller rectangles, with their points determined in the same manner. These smaller rectangles representing the irregular portions of the project can be laid out either inside or outside of the primary rectangle, dependent upon their location in the plans as illustrated in Fig. 4-8. This process will require a greater number of sightings and a higher degree of care, and will probably result in an increased error rate. For these reasons, be sure to take enough proving sightings to ensure that the work is accurate.

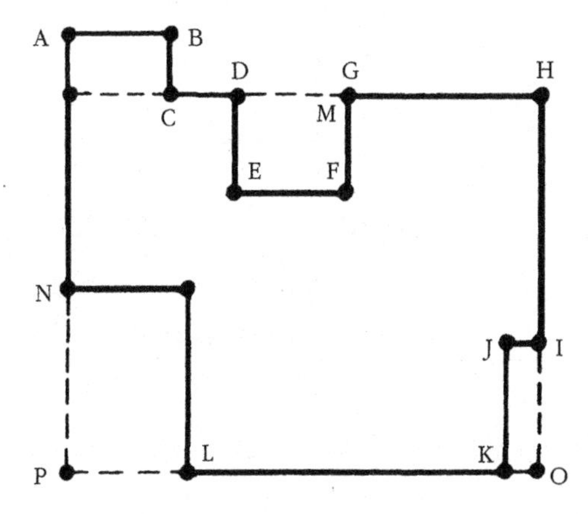

4-8 *Adapting irregularly shaped structures for plotting.*

Dimensions of excavation

There are many variables to consider when removing earth to create a sub-grade space. One of the first considerations is to consult the plans to determine the type of sub-grade space: a crawl space, a half basement, a full basement, an office space, a storage space, a living space, or a mechanicals room (plus others). For purposes of this example, assume the sub-grade space is to be a full basement, intended for multi-purpose, non-living space use.

Details of the excavation will be taken from a wall section view of the plans similar to the view shown in Fig. 4-9. Normally, the plans will dictate guide lines for the excavation, usu-

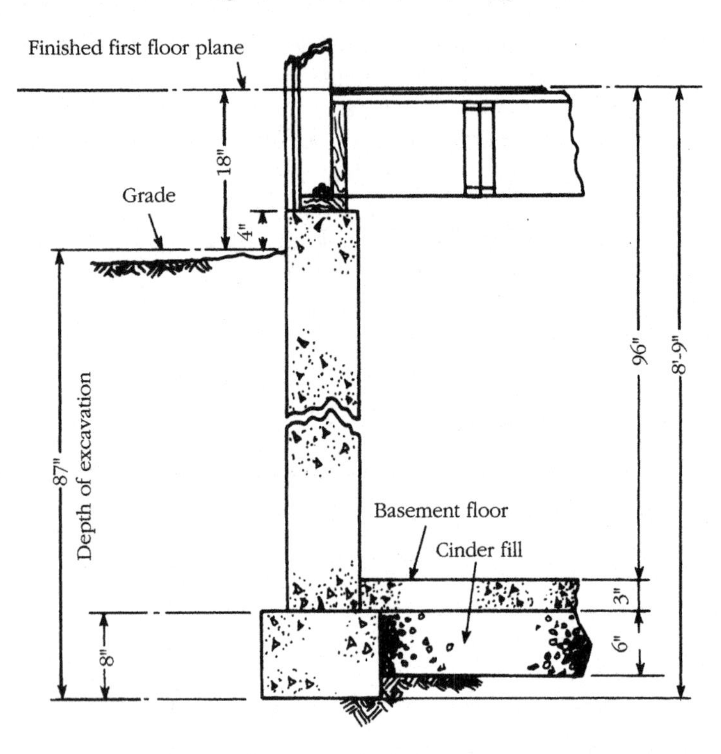

4-9 Excavation details including a prescribed first floor elevation.

ally with words similar to the following: "Excavations shall extend 2 feet 0 inches outside of all basement wall planes and to 9 inches below finished planes of basement floor levels." These particular instructions are very common and allow for two very important considerations. A 2-foot working space on the outside of the foundation is necessary to accommodate tradespeople who need to perform work on the foundation itself, such as weatherproofing. The other consideration is allowing for appropriate fill (usually 6 inches) to be put down on the surface, on top of which will be the 3-inch basement flooring.

To determine the depth of the excavation, add the vertical distance between the finished basement floor surface and the first floor finished planes, plus the 9 inches below the finished basement floor surface (3 inches of floor and 6 inches of fill), minus any vertical distance between the surface grade and the first floor finished plane (1 foot 6 inches here). The math would be 8' + 9" − 1'6".

After converting all measures to inches for ease of calculation (8' = 96" and 1'6" = 18"), the equation is now as follows:

$$96'' + 9'' = 105''$$

$$105'' - 18'' = 87''$$

$$\frac{87''}{12''} = 7'3''$$

So the base excavation would be carried to a depth of 7 feet 3 inches below grade. Note, however, that in Fig. 4-9, the top of the footing comes parallel to the surface of the fill, and that the bottom of the footer itself is below the bed of the fill. Consequently the footers would require additional excavation (normally 2 inches).

If the finished first floor elevation is specifically dictated in the plans, then the depth of the excavation would be calculated without regard to the surface grade. Referring to Fig. 4-9, if an elevation of 150 is given for the first floor, simply subtract 8 feet 9 inches from 150 to arrive at an elevation of 141.1.

In all excavations (even the shallow ones) it is necessary to either properly shore up the sides of the excavation or lay back the bank in order to prevent cave-ins. The cost of material and labor to perform this preventative work will be far less than the expense of digging out an excavation after a cave-in or slide. Use appropriate bracing or sheathing that is evenly spaced and secured along the sides of the excavation. Or, in lieu of bracing, slope the sides to the angle of repose (from 20 to 45 degrees in most instances). The angle of repose in simple terms actually refers to an angle at which the earth rests upon itself. In the laboratory, the angle of repose is determined by pouring granular material through a funnel (making sure to cause as little impact as possible) allowing the granules to pile up. These granules are to be representative of the type of earth being excavated. Once the pile is formed the angle of the natural slope of the side of the pile is measured from the horizontal, this measurement will be the angle of repose.

Since the composition of earth differs from plot to plot (even adjacent plots can differ in composition) the excavation team will need to exercise their judgment as to the angle of slope at which to lay back the banks. As stated earlier, 20 to 45 degrees will be sufficient in most cases, which corresponds to a vertical slope of 2.8 : 1 to 1 : 1. If the ground being excavated is very fine, a gentler slope may be required to avoid the soil from mucking up and slumping during a rain.

Costs of excavation

Unlike years ago, when excavations were dug with picks and shovels, modern excavations are done primarily with heavy machinery. Although this machinery will reduce labor costs by performing the work of many laborers in much less time, there are some downsides: the expense of the machine itself, trained operators, storage facilities, trucks and trailers to provide transportation of the machinery to and from work sites, annual license fees for vehicles, maintenance personnel, parts for repairs, fuel, and oil. All these things must be figured into the total cost.

Rent or buy

The question of whether to rent or lease a piece of equipment, or to purchase the equipment outright is not an easy one to answer. There are certain tax benefits inherent with each option, and the individual circumstances of the individual contractor will, in the end, dictate if leasing or buying is more beneficial. One of the benefits of short-term equipment rental is that the contractor can try out a piece of equipment prior to making a long-term lease or purchase decision.

The cost of money

When making a decision concerning the leasing or buying of equipment, financial calculations come into play. Several questions must be answered:

1. Should cash be paid for the purchase of the new equipment?
2. Should the new equipment be leased, and if so, for how long?
3. Should a loan be taken out to cover the cost of the new equipment?

Paying cash

Assuming that current cash reserves of the business would permit the outright purchase of the new equipment, the contractor must determine if utilizing the cash in this manner is the most financially sound decision. Normally, it is always less expensive to purchase the item(s) for cash rather than using borrowed funds. The exception is a situation that would allow the cash to be placed in an earnings position (invested), resulting in income greater than the interest cost of the loan. An example of this situation would be (as it was at different times during the 1980s) monies invested in certificates of deposit (CDs), which had a locked-in rate of return of 12 percent, and then, during the term of the CD, interest rates fell dramatically, resulting in loan rates running 7 to 10 percent. Under these circumstances, borrowing money would actually make money for the borrower.

Interest rates

Determining the cost of money, as used here, actually refers to how much money (interest) it will cost to use the borrowed funds over the term of the loan.

When analyzing a financial matter, whether a purchase agreement or a lease agreement, it is important to recognize which interest rate is being addressed. Interest rates can be described in three distinct ways: the periodic rate, the annual nominal rate, and the annual effective rate. For the purpose of illustration, assuming a loan of $1000, the three different rates would break down as follows:

1. *Periodic rate*: The rate of interest applied to the loan one period at a time. Periods can be monthly, quarterly, semi-annually or annually.
2. *Annual nominal rate*: The periodic rate multiplied by the number of periods in a year. If the monthly periodic rate is ½ percent, the annual nominal rate would be 6 percent (½ × 12 = 6).
3. *Annual effective rate*: The annual rate that takes compounding into consideration and is today usually referred to as the Annual Percentage Rate (APR).

Periodic and nominal annual rates can be reflective of either simple interest, or compound interest.

Simple interest

Simple interest can be illustrated by the example of one friend loaning money to another friend: if friend A loans friend B $1000 for a one-year period of time, with the understanding that at the end of that year friend B will return the original $1000 to friend A, plus an additional $100 as payment for the use of the money, then friend B is actually paying 10 percent simple interest (1000 × 10 percent = 100) on the loan.

Compound interest

The annual effective rate or APR corresponds to situations involving compound interest. Compound interest is most often calculated on either an annual, semi-annual, quarterly, or monthly basis. The periodic rate on contracts carrying compound interest reflects how often the interest is compounded. If the annual percentage rate is stated as 10 percent and the periodic rate is given as 2.5 percent, then the interest is being compounded quarterly (2.5 × 4 = 10). If a 10-percent APR is stated and periodic rates are 5 percent, then the interest is compounded semi-annually (5 × 2 = 10). Most often, interest is compounded on a monthly basis, so with a 10-percent APR, the periodic rate would be

$$8\tfrac{1}{3} \text{ percent} \left(\frac{10}{12} = 0.8333\right)$$

Compounding refers to the practice of periodically assessing interest on the unpaid balance of a loan. In the case of a monthly payment schedule, each payment decreases the size of the principal (the amount borrowed) by a certain amount, and the rest of the payment goes to pay the amount of interest due for that period of time. At the beginning of a loan repayment schedule, most of each payment is applied to the interest, with the principal being decreased only slightly. As the term of the loan goes on, the amount of the payment applied to the principal increases and the amount of interest due per period declines. Table 4-1 illustrates a loan schedule for a $10,000 24-month loan at 9.75 percent APR. This table presents monthly balances and the amounts applied to principal and interest from each monthly payment.

Table 4-1 Monthly balances, interest and principal payments on a $10,000 24 month 9.75% Apr loan.

Payment no.	Payment amount	Amount applied to interest	Amount applied to principal	Balance
1	460.30	81.25	379.05	9,620.95
2	460.30	78.17	382.13	9,238.82
3	460.30	75.07	385.23	8,853.60
4	460.30	71.94	388.36	8,465.24
5	460.30	68.78	391.52	8,073.72
6	460.30	65.60	394.70	7,679.02
7	460.30	62.39	397.90	7,281.12
8	460.30	59.16	401.14	6,879.98
9	460.30	55.90	404.40	6,475.58
10	460.30	52.61	407.68	6,067.90
11	460.30	49.30	411.00	5,656.91
12	460.30	45.96	414.33	5,242.57
13	460.30	42.60	417.70	4,824.87
14	460.30	39.20	421.09	4,403.78
15	460.30	35.78	424.52	3,979.26
16	460.30	32.33	427.96	3,551.30
17	460.30	28.85	431.44	3,119.86
18	460.30	25.35	434.95	2,684.91
19	460.30	21.81	438.48	2,246.43
20	460.30	18.25	442.04	1,804.39
21	460.30	14.66	445.63	1,358.75
22	460.30	11.04	449.26	909.49
23	460.30	7.39	452.90	456.58
24	460.30	3.71	456.58	-0-

Using the previous example of a $1000 loan at 10-percent interest compounded monthly for a one-year term, the amount repaid at the end of the year will come to $1104.71. This corresponds to an effective annual interest rate of 10.4713 percent instead of the 10-percent simple interest discussed earlier; the higher rate is due to compounding.

To determine different aspects of compound interest bearing loans, the following formulas are used:

$$A = P \times (1 + i)^n$$

where

A = the total repayment amount of the loan
P = the principal (the amount borrowed)
i = the periodic rate of interest
n = the number of periods in the term of the loan (expressed as an exponent).

Note: Remember that operations within parenthesis are calculated first.

Using the same example, A would be the unknown; P = 1000; i = 0.008333 (0.10 / 12); and n would be 12. The calculations would look like this:

$$A = 1000\ (1 + 0.008333)^{12}$$

Performing the calculations within the parenthesis first equals:

$$1 + 0.008333 = 1.00833$$

$$1.00833^{12} = 1.104709$$

The remaining equation now reads:

$$A = 1000 \times 1.104709$$

$$A = 1104.7090$$

Rounding up for dollars and cents results in:

$$A = 1104.71$$

In order to find the total amount of interest that would be paid on this loan, subtract P (the amount borrowed) from A:

$$1104.71 - 1000 = 104.71$$

To determine the effective annual rate of this loan, remember that interest equals principal times rate times time. By using the components of this equation and the answers arrived at previously, the rate can be calculated by dividing the amount of interest by the principal. In this example:

$$\frac{104.71}{1000} = 0.104710$$

In other words, the APR equals 10.47 percent.

Leasing

Leases are contracts designed to provide financing to the customer for the majority of the useful life of the item(s) being leased. Some leases include a clause entitling the customer to eventually purchase the item(s). These types of leases are referred to as capital leases, as opposed to an operating lease (which is another name for a rental agreement). The major difference between these two types of leases is that with a capital lease, most of the risks of ownership is transferred to the customer, even though legal title to the item(s) does not.

In most cases, where the equipment to be purchased (earth-moving equipment, in this instance) carries a considerable price tag, capital leasing is a popular way of financing this expense. Generally, an appropriate interest rate is applied to the periodic lease payments to compensate the lessor (the party owning the equipment) for the use of his money, which is currently in the form of the equipment the contractor is leasing. These interest rates, periodic, annual nominal, and annual effective (APR) are calculated the same as with a loan.

Differences between taking out a loan to buy a piece of equipment and leasing the same equipment are basically few. With a lease, the first payment is usually required at the time the lease is signed and, depending on state and Federal laws, the interest rate on leases may vary from those permissible on loans.

Financial functions of computers and calculators

Once again, computers and calculators prove themselves to be valuable tools. Many calculators today have the capacity to perform complex financial calculations, including figuring out loan schedules, payments, and amortization. They will convert nominal interest rates into effective interest rates and simplify each of the procedures shown in this chapter. Personal computers, as mentioned earlier in this book, are capable of many wonderful feats. With most average financial software packages today, the desktop personal computer can prepare highly detailed financial reports, loan schedules, lease schedules, mortgages, accounting functions, and more. Once again, all of the computations in this chapter can be performed easily and quickly on a computer.

One more benefit contractors can derive from the computer is that with the aid of a modem (a device that allows your computer to use the phone) shopping for equipment becomes much easier. A modem allows a computer to "call" anywhere in the world on the internet, including businesses that specialize in the sale and lease of new and used construction equipment. Other businesses on the internet are construction specialists who sell replacement parts for equipment, still others are construction consultants available to assist with any phase of a project.

Utilizing the power of a computer to locate a used piece of equipment could result in substantial savings when compared to buying new equipment.

Review questions

1. A land company has the opportunity to purchase 76.80 acres of land in Boise, near the Cole/I-84 Interchange, for $20,000 an acre. The property is zoned for light industrial and retail use.

   Developed property of this sort is currently selling for $3.50 a square foot. As configured, it appears that 415,000 square feet of the land will need to be dedicated as non-sellable roadways, sidewalks, and median strips.

   Engineering fees and improvements (including sewer lines, telephone, electrical power, curbs, and gutters) will cost the company an additional $1,250,000.00.

   To receive an adequate profit after salaries, interest charges, realtors' fees, and other fixed costs, the land company must receive a markup of at least 150% above the dollar amount of its purchase and development costs (250% of the total cost of the development.)

   Commercial lots are generally sold on a square footage basis.

a. What is the minimum price the land company must receive, on a square footage basis, for the land in this subdivision? (Hint: An acre contains 43,560 square feet.)
b. Does it make economic sense for the land company to purchase the property in question 1? Support your conclusion.

2. An important customer has asked you to bid on a construction project. You estimate labor costs at $100,000; materials and supplies at $50,000; permits and related costs at $10,000; overhead at $20,000; and taxes at $30,000.

   You know from experience that you should project a contingency of 15% for cost overruns.

   a. How much should you bid to complete the project and make a 20% profit?
   b. What percent profit would you make if no cost overruns are experienced?

3. A builder needs to place a house on a lot that has no parallel sides. How would he place the house on the lot and be sure the house is square?

5

Math for concrete & other masonry work

Many consider masonry work to be an art form, and as artists throughout history have discovered, mathematics plays a crucial role in the successful expression of the art.

Concrete work

Concrete is not concrete unless all four of its components (cement, fine aggregate, coarse aggregate, and water) are present. The quality of the concrete depends on how these four components were proportioned and mixed. Of the components that compose concrete, the ratio of cement to water is the most crucial. Because concrete can be mixed to suit the needs of many different projects, the proportions of the four components change to accommodate the particular strength of concrete required for a particular purpose. However, the ratio between water and cement used to mix concrete to be used for any particular purpose is a finite formula. The correct ratio of water to cement will result in the desired strength of the concrete.

Because of water's crucial impact on the ultimate strength of the concrete, cement sacks usually have directions giving several optional amounts of mixing water to be used per sack, which will result in several different strengths of concrete. These different mixtures are commonly called *pastes* and are usually stated as a 5-gallon paste, 6-gallon paste, or 7-gallon paste. The type of paste (or strength of concrete) required will depend on the application.

Selecting concrete & mix proportions

With any concrete application, the components of the concrete should be clean (i.e., free of foreign material). Water used for mixing should be free of any dirt, acids, rust, heavy mineral content, or detergents. Coarse aggregate needs to be hard and durable, normally made of gravel or crushed stone. The size of the coarse aggregate can vary with the type of application in which it is to be used. As a rule of thumb, the largest coarse aggregate should not exceed 20 percent (⅕) of the finished wall thickness of a normal foundation. If mixing concrete for use in a slab, the largest coarse aggregate should not exceed ⅓ (roughly 33 percent) of the slab's thickness. Coarse aggregate up to 1½ inches in size may be fine for use in concrete intended for extra thick foundation walls or unusually heavy footers.

Fine aggregate will normally be sand of uniformly fine size, with particles no bigger than ¼ inch. An important consideration when selecting sand as a fine aggregate is its moisture content. Sand (and other materials used as fine aggregate) will almost always contain some water, but the moisture content of the fine aggregate will directly affect the amount of water to be used in the mixture of the concrete. Without clinically measuring the amount of water in the fine aggregate, its moisture content can be categorized as very wet, wet (average moisture content of sand), damp, or dry. The contractor can determine which of these categories the fine aggregate falls into by squeezing a sample in her hand. If, when the hand is opened, the sand sparkles and leaves the hand wet, the aggregate is very wet; if the sand forms a ball which holds its shape, the aggregate is wet (average); should the sand crumble and fall apart upon opening the hand, the aggregate is damp; and if the sand is complete dry, with no compaction of the particles at all, the aggregate is dry (a very rare condition for aggregate to be found in).

Involving the condition of the fine aggregate, as discussed earlier, Tables 5-1 through 5-3 present the appropriate propor-

Table 5-1 Tabular method of estimating materials in a 1 : 2 : 2¼ mix.

Thickness in inches	Cubic yards of concrete	Sacks of cement	Cubic feet of fine aggregate	Cubic feet of coarse aggregate
3	0.92	7.1	14.3	16.1
4	1.24	9.6	19.2	21.7
5	1.56	12.1	24.2	27.3
6	1.85	14.3	28.7	32.4
8	2.46	19.1	38.1	43.0
10	3.08	23.9	47.7	53.9
12	3.70	28.7	57.3	65.7

Table 5-2 Tabular method of estimating materials in a 1 : 2½ : 3½ mix.

Thickness in inches	Cubic yards of concrete	Number of cement sacks	Cubic feet of fine aggregate	Cubic feet of coarse aggregate
3	0.92	5.5	13.8	19.3
4	1.24	7.4	18.6	26.0
5	1.56	9.4	23.4	32.8
6	1.85	11.1	27.8	38.9
8	2.46	14.8	36.9	51.7
10	3.08	18.5	46.2	64.7
12	3.70	22.2	55.5	77.7

Table 5-3 Tabular method of estimating materials in a 1 : 3 : 4 mix.

Thickness of concrete in inches	Cubic yards of concrete	No. of sacks of concrete	Cubic feet of fine aggregate	Cubic feet of fine aggregate
3	0.92	4.6	13.8	18.4
4	1.24	6.2	18.6	24.8
5	1.56	7.8	23.4	31.2
6	1.85	9.3	27.8	37.0
8	2.46	12.3	36.9	49.3
10	3.08	15.4	46.2	61.6
12	3.70	18.5	55.5	74.0

tions of concrete's four components in order to obtain the mix required for different applications.

Depending on the application, the actual grading of the aggregate and the desires of the contractor, the concrete mix may be adjusted to an easily workable state of plasticity. Remember, the ratio of water to cement is fixed; consequently the mixture may only be adjusted by adding or subtracting aggregate.

A mixture is workable when it has a consistency and plasticity allowing it to be placed into forms easily, creating a dense concrete with only minor spading and tamping. The mixture should hold the coarse aggregate in suspension and completely fill all cavities between the coarse aggregate, while maintaining a smooth plastic texture.

Estimating materials

When estimating materials for concrete work, different approaches may be taken; the tabular method, the proportional method, and the formulated rules of the Portland Cement Association are three to be covered in this section.

Tabular estimating

The information provided in Table 5-1, Table 5-2, and Table 5-3 represents approximated quantities of materials that would be required for project areas of 100 square feet, at varying thicknesses and mixed at different proportions. Because the information derived from the tabular method of estimating is approximate in nature, and in order to allow for a waste factor, you should increase all quantities by 10 percent.

Proportional estimating

One rough estimating procedure used by contractors for years is known as the *3/2s rule*. The rule refers to an assumption stating that the combined components of a concrete mix, for any given volume of concrete, will be 3/2s (1½ times) the volume of the concrete pour. This can be explained rather easily since, as mentioned earlier, the concrete mix is to be such that the space between the pieces of coarse aggregate are to be filled with particles of fine aggregate covered by the cement. Likewise, the space between the particles of fine aggregate are filled by the cement. With the spaces filled in this manner, the total volume of the mixed concrete components will actually be less than the sum of their individual volumes.

Normally, under the 3/2 rule, a ratio of 6 gallons per sack of cement is used to ensure the water-tightness of the concrete. The ratio of the dry components is usually taken as 1 : 2 : 3 (read: 1 to 2 to 3), referring to one part cement, two parts fine aggregate, and three parts coarse aggregate. Remember to allow extra water for wetting down the forms and subgrade, curing the concrete and clean up needs, as well as allowing the normal 10-percent waste factor for the cement and fine and coarse aggregates.

As an example, assume a 15-foot by 20-foot by 3-inch patio is to be poured. Using the 1 : 2 : 3 mix ratio, the process is as follows:

1. Volume of concrete equals: 15 feet × 20 feet × 3 inches = *X* cubic yds.
   a. Change the left-hand side into inches by multiplying 15 feet and 20 feet by 12: 180 inches × 240 inches × 3 inches = *X* cubic yds.
   b. This makes the left-hand side equal to = 129,600 cubic inches.
   c. Because a cubic yard equals 3 feet × 3 feet × 3 feet or, stated in inches, 36 × 36 × 36 = 46,656 cu. inches, 129,600 / 46,656 = *X* cubic yards, or 2.78 cubic yards.

   Alternately, another way of deriving the volume of concrete needed is as follows:

   a. 15 feet × 20 feet × 3 inches = *X* cubic yds.
   b. Convert 3 inches into a fraction (or decimal) of a foot: 15 feet × 20 feet × ¼ ft (or 0.25 feet), giving 75 cubic feet = *X* cubic yds.
   c. Because a cubic yard = 27 cubic feet, 75 / 27 = 2.78 cubic yards.
2. Calculate and add the 10-percent waste factor to the number of cubic yards of concrete needed: 2.78 × 1.10 = 0.28 cubic yds, so 2.78 + 0.28 = 3.06 cu. yds.

3. Apply the 3/2s rule by multiplying the number of cubic yards of concrete needed by 3/2 to arrive at the total volume.
4. Figuring out the required volume of each component refers back to the 1 : 2 : 3 mix ratio. Adding the elements of the ratio, 1 + 2 + 3 = 6; there are six parts to this mix:

   1 part cement = 1/6 of the total volume, or 1/6 × 4.59 = 0.77 cubic yds.
   2 parts fine aggregate = 2/6 of the total volume, or 2/6 × 4.59 = 1.52 cubic yds.
   3 parts coarse aggregate = 3/6 of the total volume, or 3/6 × 4.59 = 2.30 cubic yds.

Portland cement association rules

Several mathematical formulas (referred to as rules 38, 41 and 42) have been developed by The Portland Cement Association. In fact, the numbering of these formulas (or rules) are for convenience and reference and do not necessarily follow 37 previous rules. Although these formulas do not return absolutely accurate information regarding the volume of materials required on large projects, they are useful as "rule of thumb" tools to estimate the needed amounts of materials.

Rule 38: This rule is applied in conjunction with the mixing of mortar and is based on the fact that it takes approximately 38 cubic feet of raw materials to make 1 cubic yard of mortar. Assuming a mix ratio of 1 : 3, the procedure is as follows:

1. To determine the amount of cement required (expressed in cubic feet):
   a. Add the elements of the ratio: 1 + 3 = 4
   b. Divide 38 by the sum of the elements of the ratio 38 / 4 = 9.5, which gives 9.5 cubic feet of cement.
2. To determine the amount of fine aggregate required (expressed in cubic feet):
   a. Multiply the product of the first calculation (9.5 cubic feet) by the second element of the ratio (in this example 3): 9.5 × 3 = 28.5
   b. The mix will require 28.5 cubic feet of fine aggregate.

In order to prove the calculations add the two results: 9.5 + 28.5 = 38

Rule 41: Used in determining the volume of materials required in a concrete mix, this rule applies to mixes utilizing coarse aggregate 1 inch or less in size.

For example: assume a ratio of 1 : 2 : 4 where 1 (the first element of the ratio) represents the cement to be used in the mix, 2 (the second element) represents the fine aggregate and 4 (the third element) represents coarse 1 inch aggregate. The following procedure would be used, applying rule 41, to determine the necessary volume requirements.

1. Add the elements of the ratio together: 1 + 2 + 4 = 7.
2. Divide the rule number (41) by the sum of the elements: 41 / 7 = 5.86.
   So 5.86 sacks of cement will be required for this mix.
   Breaking down the mathematics of the ratio into English, remember the meaning of the ratio: For every 1 part of cement in the mix, 2 parts of fine aggregate and 4 parts of coarse aggregate are to be used. To calculate the other components needed, do the following:
3. Multiply the second element of the ratio by the number of sacks of cement required: 5.86 × 2 = 11.71 cubic feet of fine aggregate.
4. Multiply the third element of the ratio by the number of sacks of cement required: 5.86 × 4 = 23.43 cubic feet of coarse aggregate.

With rule 41 and rule 42, as in rule 38, add the results to prove the calculations. In this instance 5.86 + 11.71 + 23.43 = 41, thus proving the calculations were done correctly.

Rule 42: Also used in determining the amount of materials needed for concrete, this rule differs from rule 41 only in size of the coarse aggregate used. Rule 42 is applied when the size of the coarse aggregate to be used in the mix is between 1 inch and 2½ inches in size.

The major mathematical difference between rule 41 and rule 42 comes back in Step #2. With rule 42, instead of dividing 41 by the sum of the elements of the ratio (7), 42 is divided by 7, changing the number of sacks of cement to be used from

5.86 to 6. This change is reflected in all of the steps following Step #2 which, in this case, would change the results to:

$$42 / 7 = 6 \text{ sacks of cement}$$
$$6 \times 2 = 12 \text{ cubic feet of fine aggregate}$$
$$6 \times 4 = 24 \text{ cubic feet of coarse aggregate}$$

Again, to prove the work, add the results: $6 + 12 + 24 = 42$.

At times, it will be necessary to quote required volumes of material as weight of materials. Table 5-4 provides easy conversion rates for calculating the approximate weights of these materials.

Table 5-4 Conversion of concrete component volumes to approximate weights.

Concrete component	Volume	Approximate weight
Cement	1 cubic foot	94 pounds
Fine aggregate (dry)	1 cubic foot	105 pounds
Coarse aggregate	1 cubic foot	100 pounds

Estimating ready-prepared concrete

Estimating the volume of ready-prepared concrete involves the same measuring and math skills previously discussed. For foundations, remember that the area for walls is determined by multiplying length by height and the result multiplied by the thickness of the wall will equal the cubic measurement required.

Irregular shapes can be broken down into multiple regular shapes, measured separately and the results added together to arrive at a total figure.

Labor, form lumber, & other expenses

Normally, the single most expensive component of any job will be the labor costs. As mentioned earlier in this book, labor costs are not just the wages paid to the employees, even though that is a large part of it. Labor costs are also comprised of any state, local, or federal taxes to be paid or withheld; unemployment insurance contributions; and any medical, life, and health insurance programs that require employer contributions.

When figuring out labor cost as part of the projects' expense, do not neglect to include the portion of the office staff's and support staff's hours that can be attributed to that particular job.

In addition to labor costs, each project requires the use of tools and equipment. Floats, trowels, reinforcing rod, mixers, earth movers, compactors, shovels, and rakes are just some of the tools and equipment regularly used. Whether purchased or rented or leased, the contractor will definitely incur an expense for each of these items—an expense that must be considered by the contractor when bidding the job and figuring the profit margin on each job.

One unique expense when dealing with concrete work is that of form lumber. Most projects will require the use of forms made from lumber or plywood or prefabricated out of aluminum or steel. On a job of any appreciable size, quite a few pieces of lumber will be needed to be cleated together to attain the proper widths. As seen in Fig. 5-1, this sheathing, formed by the cleated together lumber, is then held in place by studs and braces (usually made out of 2 × 4 lumber) placed roughly every 18 inches around the outside of each form. Assuming a project that requires a foundation 19 feet

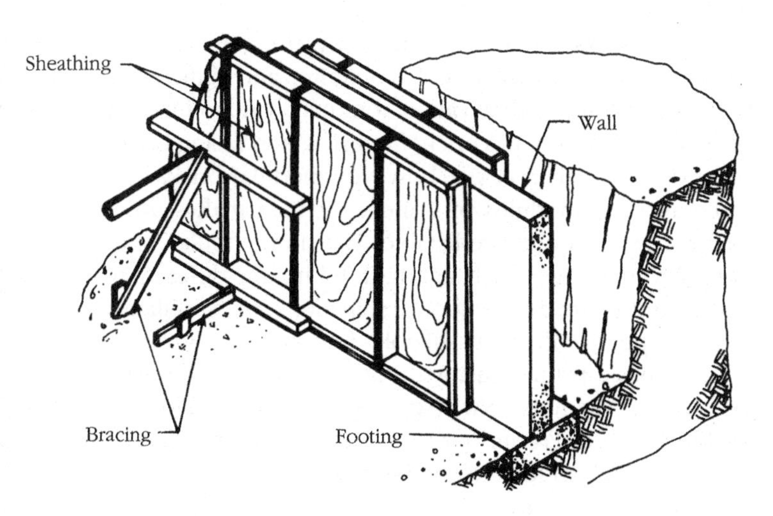

5-1 *Sheathing and bracing a foundation wall.*

square: a bill of materials for the lumber needed for the forms alone would read as follows:

Sheathing: 40 pieces, 1 inches × 6 inches × 10 feet
Plates: 16 pieces, 2 inches × 4 inches × 10 feet
Studs: 56 pieces, 2 inches × 4 inches × 2 feet
Stakes: 56 pieces, 2 inches × 4 inches × 1 feet 6 inches
Braces: 56 pieces, 1 inches × 4 inches × 3 feet
Braces: 56 pieces, 1 inches × 4 inches × 2 feet 6 inches

Added up, the total board feet of lumber needed would be 552. In addition to the expense of the lumber, fasteners and lubricant must also be figured into the cost of these forms.

Accounting for every aspect of each job will ensure a more accurate price for the job and prevent needless losses from being incurred.

Brick masonry

Estimating the number of bricks needed for a project requires the contractor to consider many different factors, not the least of which is the total area to be bricked. The thickness of the area and the joint size stipulated for the project play a major role in determining the number of bricks required to complete the project.

Most bricks are manufactured by firing molded clay or shale and vary widely in color, texture, and dimensions. All bricks can be assigned to one of four main categories: common or building bricks, patio brick, fire brick, and facing brick. Two other designations do exist, however: modular bricks are sized to conform to modules of 4 inches, which makes estimating material needs quite predictable, but non-modular bricks do not conform to the 4-inch standard. Within the different groupings of bricks are a variety of designations that stipulate a particular type of brick as suitable for specific purposes. For outdoor projects that must withstand moisture and freeze-thaw cycles, for example, SW (severe weathering) grade bricks should be used. For indoor uses, such as facing a fireplace or a planter, MW (moderate weathering) or NW (no weathering) bricks are appropriate.

When a brick's dimensions are quoted, its nominal sizes are the actual measurements being referenced. Nominal size is the actual size of the brick, plus a normal mortar joint of ⅜ to ½ inch on the bottom and at one end. With the numerous styles and sizes of brick available today, the most common brick still measures 2¼ inches × 3¾ inches × 8 inches, with an exposure of 2¼ × 8. Allowing for a ¼ inch joint, these common bricks have a total thickness of 4 inches (3¾ + ¼ = 4).

With a face exposure (including the joint) of 2½ inches (2¼ inches of brick + ¼ inch of joint) × 8¼ inches (8 inches of brick + ¼ inch of joint), each brick has an area of 20.63 square inches. For every square foot of area to be bricked there are 144 square inches (12 inches × 12 inches). To determine how many bricks will be needed per square foot of area, divide 144 by 20.63, which mathematically equals 6.98. For practical purposes this is certainly close enough to be called seven bricks.

In the case of a wall to be built with standard brick with standard ¼ inch mortar joints, multiplying the number of square feet to be bricked by 7 will result in the number of bricks needed for the construction of a 4-inch thick wall (the thickness of one brick plus its joints). If the area to be bricked requires an 8-inch thick wall, multiply by 14 instead of 7; for a 12-inch thick wall, multiply by 21. Naturally, if larger joints are used, the number of bricks per square foot will decrease. With ½-inch joints, for instance, the number of bricks per square foot drops from 7 to 6.15; with joints of ¾ of an inch, the figure drops to 5½. In addition to calculating the number of bricks required for the square footage of the wall, bricks must also be added to the count for header courses if they are stipulated. Once the total number of bricks required for the project is determined, add 5 percent as a waste factor.

Walls laid up with a 4-inch thick face brick veneer over common brick are not uncommon. If the bricks are to be laid in a common bond as in Fig. 5-2, every sixth course is usually a header course made up of face bricks and will require twice as many face bricks as a stretcher course. This type of bond will result in an overall increase of the number of face brick needed of approximately 16⅔ percent and conversely, an equal decrease in the number of common brick.

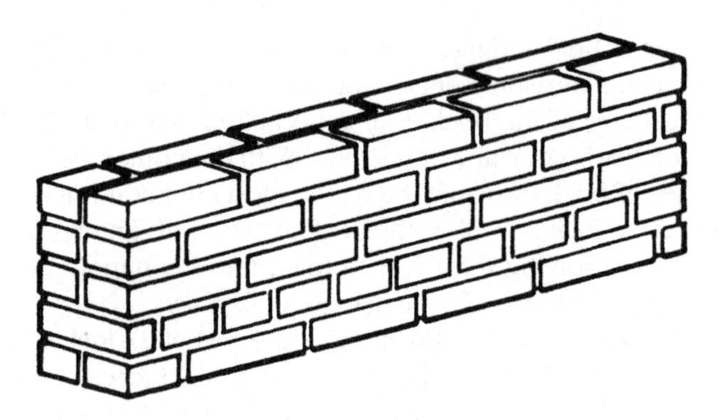

5-2 Common bond pattern of brick.

Tabular estimating

Estimating the amounts of materials needed to brick given areas can also be approximated with the use of tables. For example, suppose a 100-square-foot solid sub-grade wall with ½ inch joints is to be constructed, with the outer 4 inches laid up, with all joints filled and all other brick to be laid end to end on full mortar beds. Assuming a header course is to be laid every five courses, with the vertical spaces between the 4-inch thicknesses filled with mortar, Table 5-5 will approximate the necessary materials.

Table 5-5 Estimate of materials for sub-grade ½-inch joint brick wall at different thicknesses.

Wall thickness in inches	Number of bricks required	Amount of mortar required in cubic feet
8	1,271	19.5
12	1,926	31.4
16	2,580	43.3

For the same wall constructed above grade, with the vertical spaces remaining open, Table 5-6 can be used as a guide.

Estimating materials for either of these walls, regardless of total area, can be accomplished with the information presented in these last two tables. Algebraic formulas involving many

Table 5-6 Estimate of materials for above-grade ½-inch joint brick wall at different thicknesses.

Wall thickness in inches	Number of bricks required	Amount of mortar needed in cubic feet
8	1,271	13.5
12	1,925	19.5
16	2,580	25.5

steps can be applied to compute these estimates; however, a more simple common sense approach can also be used to complete the necessary estimate. Suppose instead of 100 square feet of wall space, a 20-foot section of 8-inch thick wall is to be constructed under the specifications given for Table 5-6. The table indicates that, for 100 square feet of wall, 1271 bricks will be needed. To calculate how many bricks are needed for 20 feet of wall, first divide 1271 by 100:

$$\frac{1271}{100} = 12.71$$

It is now known that there are 12.71 bricks per square foot. Next multiply the number of bricks per square foot by the number of square feet to be constructed:

$$12.71 \times 20 = 254.20$$

Add 5 percent as a waste factor:

$$254.20 \times 0.05 = 12.71$$
$$254.20 + 12.71 = 266.91$$

Rounding up to an even number determines that 267 bricks will be needed to complete the proposed project.

In the same common-sense manner, if a 130-foot wall is to be similarly constructed, the following steps would be followed:

1. Subtract 100 from the total square footage to be built: 130 − 100 = 30.
2. Determine the number of bricks required per square foot: 1271 / 100 = 12.71.

3. Multiply the square footage in excess of 100 (from Step #1), in this case 30, by the number of bricks needed per square foot: 30 × 12.71 = 381.30.
4. Add the results of Step #3 to the appropriate number of bricks for 100 square feet supplied by the table: 381.30 + 1271 = 1652.30.
5. Multiply the total from step 4 by 0.05 (5 percent) and add the result to the total from Step #4: 1652.30 × 0.05 = 82.62, so 82.62 + 1652.30 = 1734.92.

Again, rounding up to an even number of bricks results in 1735 bricks needed for a wall measuring 130 square feet.

Using the same common sense calculations as above, materials can be estimated for brick piers and footings, based on Tables 5-7 and 5-8.

Table 5-7 Estimating material requirements for solid brick piers.

Size of piers	Amount of bricks needed	Amount of mortar needed (in cubic feet)
8 inches by 12 inches by 10 feet	124	2.25
12 inches by 12 inches by 10 feet	185	3.25
12 inches by 16 inches by 10 feet	247	4.50

Table 5-8 Estimating required materials for brick footings.

Thickness of construction at 100 linear feet	Number of bricks required	Cubic feet of mortar necessary
8 inch	2,272	39
12 inch	2,812	48
16 inch	4,592	78

Adjusting materials estimates

For brick walls that are to include windows and doors, the amount of required materials must be adjusted to account for these openings. The net surface area is arrived at by simply calculating the total area of all window and door openings and subtracting this figure from the gross surface area.

For construction stipulating face brick backed by common brick or block, window and door openings in walls dictate an addition to the materials estimate. For each linear foot around the four sides of each window opening and the three sides of door openings, 1½ face bricks need to be added to the estimate. Brick veneer construction will require five additional bricks per linear foot across the top of each window and door to form a soldier course. Rowlock courses across the bottom of each window and door will also require an additional five bricks per linear foot.

Estimating labor

Much of estimating labor costs is dependent upon the contractor knowing her own capabilities or the capabilities of her individual crews. A more experienced crew may be much faster at laying up a chimney than another, or one crew may be quicker at laying up English Bond than another crew. These factors must be considered when estimating labor costs for bidding purposes. However, for use as a guide, Table 5-9 presents ranges of man-hours a professional mason would require to complete several different tasks, based on common bond construction. For any other bond, labor estimates should be increased by 10 to 15 percent. Bear in mind that these hours are for the mason and do not include man-hours which should be charged to the job for helpers.

Table 5-9 Brick-laying estimates for masons.

Type of construction to be completed	Man-hours for 1 mason per 1,000 bricks
Wall: Common brick, finished on one side	5 to 12 man-hours
Wall: Common brick, finished on two sides	6 to 14 man-hours
Wall: Face brick construction	10 to 20 man-hours
Wall: Firebrick construction	18 to 32 man-hours

Concrete block

Due to their economy and versatility of uses, concrete blocks are manufactured in a variety of sizes and shapes to more easily conform to differing construction needs. Concrete blocks are cast from a stiff concrete mix. A typical stretcher block has a nominal size of 8 × 8 × 16 inches and weighs 40 to 50 pounds. Hollow cores in the block conserve material, make them lighter in weight, easier to grip and place, provide a space in which to add insulation, and provide channels through which to run utility lines.

Concrete blocks come in two different grades: N grade, which is designed for outdoor conditions where repeated freezing and thawing cycles can easily occur; and S grade, for use above grade and where it will not be exposed directly to the weather.

Although blocks may be cut on site with a power masonry saw, it is much more economical to lay out the job plan to take advantage of the manufactured sizes and shapes. Full-length and half-length units are available, as are stretcher units, corner units, header, and bull-nose units, among many other blocks (Fig. 5-3) designed for specific purposes.

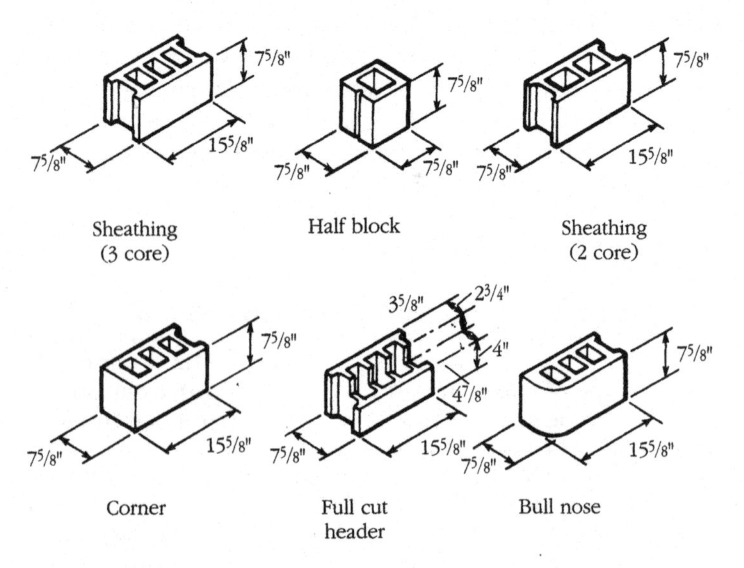

5-3 *Different types of concrete block.*

A concrete block measuring 7⅝ by 7⅝ by 15⅝ will normally be referred to as an 8-×-8-×-16-inch block. This reference to a block's nominal size is common practice and is based on the space the block will occupy when laid in place with a ⅜-inch mortar joint. Using these nominal measurements, all horizontal and vertical elements of the project should be laid out in multiples of these dimensions, utilizing both full- and half-size blocks.

Because there are many specialty blocks used in different applications throughout the construction of block walls, the number of these specialty blocks must be subtracted from the number of regular blocks required for the project. Using the take-off lists to identify all concrete blocks other than regular full- and half-size blocks, the sum is then subtracted from the total number of blocks to be used.

Table 5-10 is a guide, which will be useful in estimating the gross number of blocks and the amount of mortar needed for a project. Remember to reduce the gross amount to the net amount by subtracting the number of specialty blocks called for by the plans. As with bricks, always add 5 percent to the net total to allow for waste.

Estimating labor costs for laying block, as in many other areas, is a mix of the contractor's knowledge of her capabilities or those of her crews, the contractor's experience, and the degree of difficulty of the work to be done. Table 5-11 provides

Table 5-10 Estimating required materials for concrete block construction.

Nominal block size to be used	Number of blocks needed per 100 square feet	Block height	Mortar required per 100 square feet of area
4 by 4 by 16	225	4 inches	13.5 cubic feet
6 by 4 by 16	225	8 inches	8.5 cubic feet
8 by 4 by 16	225		
4 by 8 by 16	112.5		
6 by 8 by 16	112.5		
8 by 8 by 16	112.5		
12 by 8 by 16	112.5		

guidelines for estimating labor hours based on the average range of time it takes one mason to lay, clean, and point 100 blocks. Remember, these figures do not include any man-hours for helpers.

Table 5-11 Estimating labor costs for concrete block information.

Size of blocks to be used	Average man-hours per 100 blocks
2 by 8 by 16	3.0 to 3.5
4 by 8 by 16	3.5 to 4.0
6 by 8 by 16	4.0 to 4.5
8 by 8 by 16	4.7 to 5.2
10 by 8 by 16	6.0 to 6.5
12 by 8 by 16	7.0 to 7.5

Stone masonry

The stone used for construction generally falls into three categories: rubble (round rocks), flagstone (flat irregular pieces), and ashlar or dimensioned stone, which is cut into slices of uniform thickness for laying in coursed or non-coursed patterns. These different types of construction stone may be measured by the piece, by the ton, by volume, by area, or by weight. Regardless of how the stone is categorized, the contractor must be able to determine how much stone will be required to complete a given project.

Estimating stone requirements

One of the most popular (and dependable) methods for estimating stone quantities is by using a combination of weight and volume. If the cubic foot yield of one ton of a particular stone is known, then simple calculations will determine how many tons of stone will be needed for the project. Table 5-12 lists some of the more popular stone building materials along with their weights per cubic foot.

For demonstration purposes, imagine an 8-×-10 sandstone wall is to be laid up, and assume an average weight of 150

Table 5-12 Approximate average weight per cubic foot of building stone.

Type of building stone	Approximate average weight per cubic foot
Limestone	150 to 175
Sandstone	140 to 160
Granite	165 to 175
Marble	165 to 175
Slate	160 to 180
Dolomite	155 to 175

pounds per cubic foot of the stone. The following procedure will result in the estimate for the quantity of stone required:

1. One ton of sandstone divided by the weight per cubic foot of the stone equals the number of cubic feet per ton: 2000 / 150 = 13.33.
2. Convert 13.33 cubic feet into square feet by dividing one foot by the average thickness of the stone. If the thickness is 6 inches, then 12 inches / 6 inches = 2 inches.
3. Multiply the number of cubic feet per ton by the number derived in #2 to determine the gross square footage (wall area) per ton: 13.33 × 2 = 26.66.
4. Divide the total net area to be laid up (10 by 8 = 80 square feet) by the gross square footage per ton: 80 / 26.66 = 3.

The estimate shows that 3 tons of sandstone will be necessary for this particular project.

Estimating mortar requirements

The composition of the mortar used in stone masonry is often very similar to that used in brick masonry. Differences usually involve hydrated lime being added to the mix in an effort to improve the workability of the mortar, or using stone-set cement instead of portland in order to give a whiter appearance to the mortar.

Factors that affect the amount of mortar used for the job are the size of the stones used, the pattern in which they are laid, and the size of the joints between the stones. Table 5-13 supplies a guide for estimating the amount of mortar necessary per cubic yard using these variables.

Table 5-13 Estimated mortar requirements per cubic yard of stone work.

Type of bond and joint size	Amount of mortar required (in cubic feet)
Coarsed rubble	6½ to 8½
Coursed ashlar with ¼-inch joints	1½ to 2
Coursed ashlar with ½-inch joints	3 to 4
Random rubble	7½ to 9½
Random ashlar with ¼-inch joints	2 to 2½
Random ashlar with ½-inch joints	4 to 5
Cobblestone	6½ to 9½

Stone costs

The cost of stone varies, not only with the type of stone, but with the quantities purchased, the amount of work done by the supplier (cutting, splitting, etc.), and the region of the country where the contractor is located. If plans for a project in Texas call for Pennsylvania slate, the stone will naturally be more expensive than if the project were in Pittsburgh.

The basic components that a contractor must consider when discussing the cost of stone are as follows:

- The cost of the stone charged by the supplier
- Transportation cost of the stone to the job site
- The cost of any cutting/shaping done on site, such as end joints

Bear in mind that none of these costs include mortar or labor hours for laying up the wall but are just components of the cost of the stone itself.

Review questions

1. A cement driveway between the garage and the street, a distance of 135 feet, needs to be built. The driveway

needs to be 4 inches thick. Cement costs $55/cubic yard with a 15% discount for cash.

a. How many cubic yards of cement will be needed for a driveway with a width of 15 feet?
b. How much will the builder pay if he uses cash?

2. The Highway District Pavement Management Technician needs to know how many tons of asphalt will be needed on a section of road. The section of road measures 5280 feet (1 mile) in length and 26 feet wide. The asphalt needs to be 3 inches in depth. Asphalt weighs 144 tons per 2000 ft^3.
3. You need to pour a new sidewalk that is 4' wide, 16' long, and 4" thick. Concrete costs $70.00/$yard^3$ (placed).
   a. How many cubic yards of concrete are required?
   b. How much will the concrete cost?
4. If the length of one side of your house is 44 feet and the adjacent side is 68 feet, show that all corners form 90-degree angles.
5. How many bricks will be needed to build a wall 68 feet long by 6 feet high by four inches thick. Use standard brick with ¼ inch mortar joints.

6

Math for floor framing & covering

When framing a floor or covering a floor, one must consider weight, size, and application. Each of these attributes, plus many others involved with flooring, use mathematics to determine the necessary materials and actions needed to complete the job.

Live loads & dead loads

As construction begins on a structure, the dead load of the structure also mounts. Because the dead load is the aggregate weight of the structure itself, the dead load increases as the structure is built and continues to increase until the building is complete. Once finished, the dead load of the structure will remain consistent unless major changes are made to the structure. All structural elements of the building are part of the dead load. Window air conditioners are not considered dead load, but industrial-size air-handling equipment located on the structure's roof definitely is part of that structure's dead load. During the life of the building, the addition or deletion or relocation of heavy-duty air-handling plants to or from any area other than the concrete foundation floor will either increase or decrease the dead load of the structure.

The live load of a structure is made up of a combination of two components: the movable objects (including people) within the structure and any external (natural) forces that would be likely to periodically act upon the outside of the building. Using an office building as an example, one of the first components of the live load to be considered would include the number of employees who will be working there. The office furniture, office machines, and an anticipated number of non-employee personnel who may be present at any given time will also be components of the live load. A multi-story office complex, such as a government office complex, can very easily expect several hundred non-employee personnel (visitors) present at any given time during the day on a regular basis. The second component of the same structure's live load takes into consideration the location of the building. For example, if the structure is located within an area of the country normally receiving snowfall during winter, provisions must be made for the additional weight of accumulated snow and or ice on the building's roof.

Beams

Strategically placed beams and their support columns bear a major portion of the combined dead and live loads of a structure, as seen in Fig. 6-1. To determine the location and number of beams and columns necessary for a structure allowable joist length, room arrangement and the location of weight bearing partitions must be considered. As a general guide for residential structures, if joist span would exceed 14 feet when using single girder construction, the addition of a second girder is called for. The addition of the second girder will prove economical because it will keep joists considerably less than 14 feet in length, eliminating the need for double rows of bridging.

Calculating beam strength

The strength of a wood beam is, naturally, based on the size of the beam. However, the effects of the length, width, and depth of the beam on its strength may not be as logical as they first appear. When estimating the dimensions of a beam, consider the following:

- Think of two boards, one 10 feet long and the other 20 feet long, both with evenly distributed loads along their lengths and supported on either end. Both will tend to sag or bend, but the 20-foot board will bend more radically and be more likely to break. Because the longer board is twice the length of the shorter board, a logical first assumption is that the safe load capacity per foot of the longer board is ½ that of the shorter board. However, since an additional foot of load is added to the board for each additional foot of length, mechanical theory and experience have proven that the actual safe load capacity per foot of the longer board would be only ¼ that of the shorter board. For this reason, beam strength must be increased to offset the decreased load capacity of unsupported beam spans. The greater the unsupported span, the greater the required beam strength. Increasing the width and depth of the beam and/or using an intrinsically stronger material for the beam will effectively increase its strength.
- In the case of width, logic does hold true. Doubling the width of a beam will double its strength. However, the doubling of a beam by fastening two identical beams together will not have a higher capacity than the two beams placed separately; the load per foot capacity remains the same.
- Doubling the depth of a beam is the most effective way of increasing its strength. A beam which is 4 inches by 8 inches will support four times the load of a beam measuring 4 inches by 4 inches. Although doubling the depth of the beam is a very efficient way of strengthening it, consideration to practical applications must be given. While a beam measuring 18 inches in depth may provide the strength required, normally, any beam with a depth greater than 10 inches would not be feasible due to headroom constraints.

Determining beam size

The first step in determining beam size is to discern the placement of the beam's support columns. For this purpose, the total length of the space the beam will run between walls should

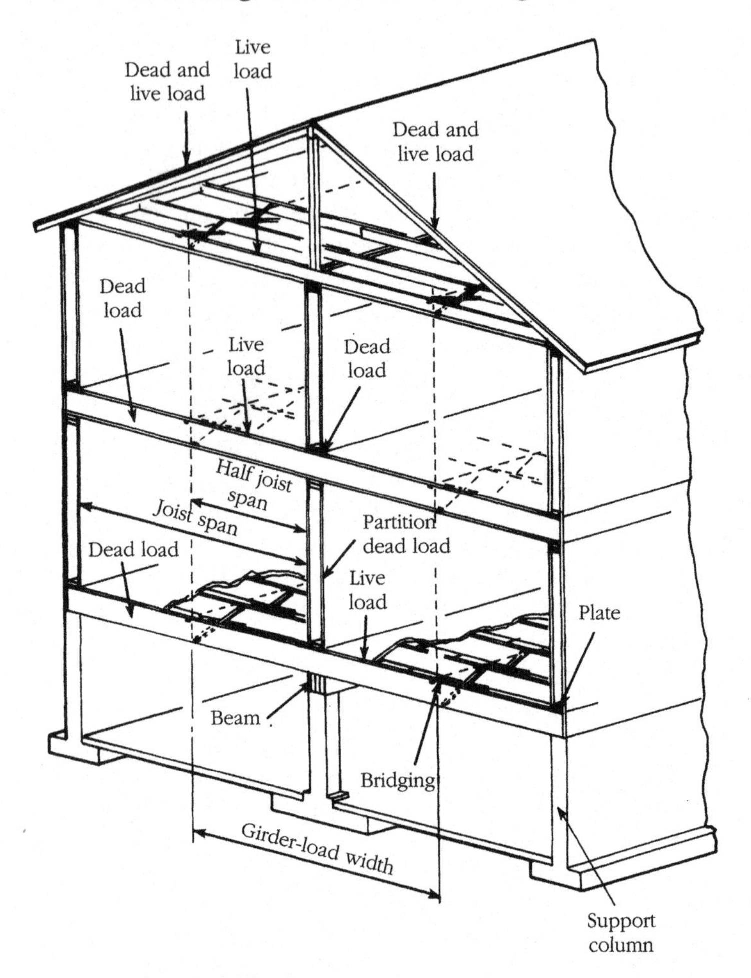

6-1 The live and dead load areas of a structure.

be divided into similarly sized segments, taking care to avoid excessively long spans.

The next indicator to be calculated is the half width. Simply stated, the half width is that portion of the joist load which the beam will carry. To calculate half widths follow this procedure:

1. From one side of the beam, measure the joist span (from the middle of the beam to the nearest joist support).
2. Determine the measurement of the corresponding span on the other side of the beam and add the two measurements together.

3. If the joists are butted or lapped over the beam, multiply the total measurement from Step #2 by ½; if the joists are continuous, multiply by ⅝ instead of ½. The results of these calculations are the half widths. Notice that continuous joists will allow the beam to support a greater proportion of the load than joist which are butted or lapped.
4. Add the dead and live load figures for each floor, the dead load figures for each floor partition, and the dead and live load figures for the roof (taken from the plans). The total of these figures equal the combined square foot floor load that the beam is to carry.
5. Multiply the combined square foot floor load figure by the half width to obtain the amount of load on the beam per linear foot.
6. Multiply the load on the beam per linear foot by the distance between beam supports to obtain the total load on the beam. For example, if the load on the beam per linear foot is 3000 pounds and the distance between support columns is 8½ feet the equation is $3000 \times 8.5 = 25{,}500$.

Which means that the total, evenly distributed load over the entire length of the beam equals approximately 25,500 pounds.

Table 6-1 presents beam loads per linear foot as they correspond to total square foot floor load. Note that the information in this table relates only to beams of even lengths. In order to calculate loads per linear foot for beams of odd lengths, subtract the figure given for the total square foot floor load on a beam with a half-width one foot shorter than the half-width of the beam in question from the figure given for a beam with a half-width one foot longer than the beam in question. Then divide this number by 2 and add the result to the figure given for a beam one foot shorter than the beam in question.

This demands an example, so let's assume a total square foot floor load of 150 and the need to arrive at the per linear foot beam load for an 11-foot beam. Now, finding the numbers on Table 6-1 relating to a square-foot floor load of 150, the em-

Table 6-1 Tabular calculation of beam load per linear foot.

Total square foot floor load	*Beam load per linear foot by half widths*										
	5	**6**	**7**	**8**	**9**	**10**	**12**	**14**	**16**	**18**	**20**
10	50	60	70	80	90	100	120	140	160	180	200
20	100	120	140	160	180	200	240	280	320	360	400
30	150	180	210	240	270	300	360	420	480	540	600
40	200	240	280	320	360	400	480	560	640	720	800
50	250	300	350	400	450	500	600	700	800	900	1000
60	300	360	420	480	540	600	720	840	960	1080	1200
70	350	420	490	560	630	700	840	980	1120	1260	1400
80	400	480	560	640	720	800	960	1120	1280	1440	1600
90	450	540	630	720	810	900	1080	1260	1440	1620	1800
100	500	600	700	800	900	1000	1200	1400	1600	1800	2000
110	550	660	770	880	990	1100	1320	1540	1760	1980	2200
120	600	720	840	960	1080	1200	1440	1680	1920	2160	2400
130	650	780	910	1040	1170	1300	1560	1820	2080	2340	2600
140	700	840	980	1120	1260	1400	1680	1960	2240	2520	2800
150	750	900	1050	1200	1350	1500	1800	2100	2400	2700	3000
160	800	960	1120	1280	1440	1600	1920	2240	2560	2880	3200
170	850	1020	1190	1360	1530	1700	2040	2380	2720	3060	3400
180	900	1080	1260	1440	1620	1800	2160	2520	2880	3240	3600
190	950	1140	1330	1520	1710	1900	2280	2660	3040	3420	3800
200	1000	1200	1400	1600	1800	2000	2400	2800	3200	3600	4000
210	1050	1260	1470	1680	1890	2100	2520	2940	3360	3780	4200
220	1100	1320	1540	1760	1980	2200	2640	3080	3520	3960	4400

phasis will be placed on the figures from 10-foot and 12-foot half widths, with calculations as follows:

$$1800 - 1500 = 300$$

$$\frac{300}{2} = 150$$

$$1500 + 150 = 1650$$

Notice that just as the total square foot floor loads are each multiples of each other, each figure representing the load per linear foot is also a multiple. For example, at a total square foot floor load of 10, the beam load per linear foot at a half-width of 5 is 50; 13 times the square foot floor load would equal 130, and 13 times the beam load of 50 per linear foot would be 650.

Consequently, with the information supplied in Table 6-1 and simple multiplication, any square foot floor load can be converted to beam load per linear foot.

At this point, the decision needs to be made whether a wood or steel beam is to be used for the application. If wood is the selection, the next step is to decide which type of wood the beam is to be made of. Table 6-2 presents information about several types of construction grade lumber and the allowable unit stress (expressed in pounds per square inch) for beams 5 inches or greater in thickness. As to the question of whether the beam should be a solid member or a built-up member, bear in mind that there are pros and cons to be considered. Built-up members, due to the difference in dressed size, will carry a marginally lesser load than will a solid member of the same approximate size. However, because creating a single beam from two or more pieces of material presents the opportunity to custom select, arrange, and fasten the material together, this process may result in a built-up beam that is more appropriate for its use than a solid beam would be.

A vertical bearing member, a column, or a post is designed to support a superimposed vertical load, which is in turn either directly or indirectly supported by the foundation. Due to the large portion of the structure's weight carried by each column or post, each member must be of an adequate size, made from appropriate material, and precisely located and seated upon a sufficient foundation.

As previously mentioned, spans between support members (columns or posts) should be limited. Usually maximum spans of 8 to 10 feet are advisable to avoid the need for extra large beams and/or placing too much weight on individual footings.

Calculating column/post size

Posts or columns, as a rule, normally carry half the weight of the beam load on either side of the support member as illustrated in Fig. 6-2, where posts A, B, and C are supporting the beam with its attached joists. Post B will support all of the weight of and on the beam from B along the beam to point E (the midway point between B and A). Likewise all the weight of and on the beam to point F (the midway point between B

Table 6-2 Allowable unit stress factors for construction grade lumber.

Type of wood	Construction grade	Allowable unit stress for beams 5 inches in thickness or greater
West Coast	Superstructural (dense)	2,000
Douglas Fir	Superstructural and dense structural	1,800
	Structural	1,600
	Common structural	1,400
Inland	Dense superstructural	2,000
Empire	Dense structural	1,800
Douglas Fir	No. 1 common dimension and timber	1,135
Western	No. 1 common dimension and timber	1,135
Larch	Select structural (extra dense)	2,300
	Select structural	2,000
	Extra dense heart	2,000
	Dense heart	1,800
	Structural square edge and sound	1,600
	No. 1 common (dense)	1,200

Southern Yellow Pine	No. 1 common dimension and timber	1,135
	Select structural (extra dense)	2,300
	Select structural	2,000
	Extra dense heart	2,000
	Dense heart	1,800
	Structural square edge and sound	1,600
	No. 1 common (dense)	1,200
Redwood	Superstructural	1,707
	Prime structural	1,494
	Select structural	1,322
	Heart structural	1,150

Allowable unit stress is expressed as pounds per inches squared.

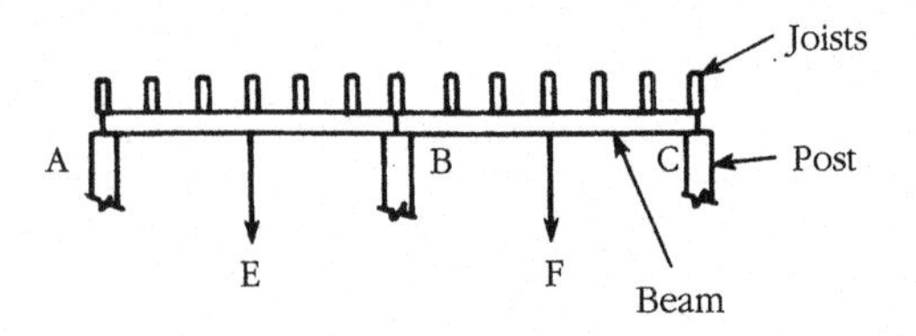

6-2 *The area of a beam's weight which a post supports.*

and C) will also be supported by post B. To calculate the actual load to be supported by a post or column, follow these steps:

1. Determine the distance (in feet) from the middle of the support to the next nearest support on one side (the span of one side).
2. Using calculations previously discussed in this chapter, multiply the beam load per linear foot by the span, in feet, arrived at in Step #1.
3. Repeat Step #1 on the other side of the beam finding its span.
4. Multiply this span by the beam load per linear foot.
5. Adding the results of Steps #2 and #4 will deliver the amount of load placed on the support.

The "but" in this procedure comes in the form of a reminder. Remember that the proportion of load carried by the support will be affected by the type of construction used. In Fig. 6-2, the beam is cut over support B, which determines that post B will carry half the load in either direction. If the beam were continuous over the support, post B would carry ⅝ of the load. Each post will need to be considered separately in each form of construction.

Another example of differing loads per post is given in Fig. 6-3, where five columns are supporting a continuous beam. The space between the two end supports is equally divided by the three interior columns. This arrangement results in the center support (post C) carrying half the load of and on the beam between itself and post B on one side and half the load between itself and post D on the other side. Post B, while supporting half the load of and on the beam between itself and

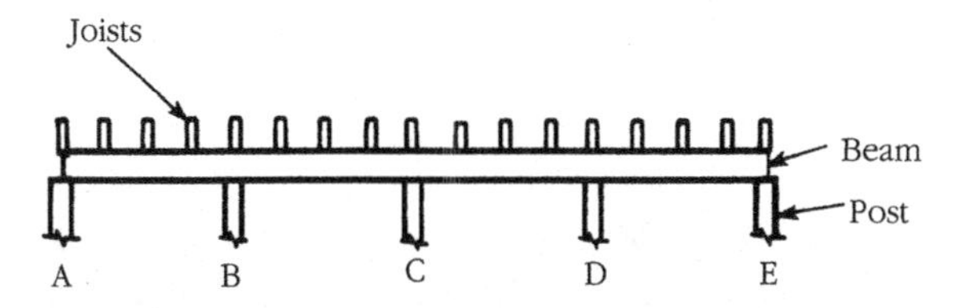

6-3 The load carried by support post in a four equal-spaced span configuration.

post C, will also handle ⅝ of the load between itself and column A. Similarly, support D will carry half the load of and on the beam between itself and C, while shouldering ⅝ the load between supports D and E.

Once the load each post is expected to support is known, two other factors affecting the actual size of each post will come into play:

1. The type of support selected for use.
2. The type of material from which the supports are to be made.

When compared to other design shapes and materials composing support columns for areas of equal size, square timber posts and round steel pipe columns have shown to provide superior strength. For ease of comparison and selection, the following tables furnish load allowances for different types of No. 1 common grade lumber frequently used in timber beams.

Like square timber posts, columns made of round steel pipe are often the preferred choice to act as support columns for beams. Similar to Table 6-3 and Table 6-4, a chart of load capacities for a variety of different diameter and lengths of steel pipe is given in the following table.

Joists

The principles involved in sizing joists are essentially the same as those used for sizing beams, with adequate strength being the primary consideration. Along with strength, however, a new element is introduced within the topic of joists: stiffness. A joist can be of adequate strength to support its load but can

Table 6-3 Maximum load capacity for support columns constructed of No. 1 Common Grade: Western Red Cedar, Spruces, White Pine, White Fir, Eastern Hemlock.

Nominal size (in.)	3 × 4	4 × 4	4 × 6	6 × 6	6 × 8	8 × 8
actual size (in.)	2⅝ × 3⅝	3⅝ × 5⅝	3⅝ × 5⅝	5½ × 5½	5½ × 7½	7½ × 7½
area	9.51	13.14	20.39	30.25	41.25	56.25
Column height (ft.)						
4	4,950 lbs.	7,280 lbs.	11,300 lbs.	16,940 lbs.	23,100 lbs.	31,500 lbs.
5	4,380 lbs.	7,100 lbs.	11,000 lbs.	16,900 lbs.	23,060 lbs.	31,500 lbs.
6	3,460 lbs.	6,650 lbs.	10,300 lbs.	16,700 lbs.	22,850 lbs.	31,500 lbs.
6½	2,960 lbs.	6,320 lbs.	9,800 lbs.	16,600 lbs.	22,700 lbs.	31,400 lbs.
7		5,960 lbs.	9,270 lbs.	16,400 lbs.	22,400 lbs.	31,300 lbs.
7½		5,630 lbs.	8,720 lbs.	1,6200 lbs.	22,100 lbs.	31,100 lbs.
8		5,160 lbs.	7,930 lbs.	15,950 lbs.	21,800 lbs.	31,000 lbs.
9		4,060 lbs.	6,300 lbs.	15,350 lbs.	20,950 lbs.	30,640 lbs.
10				14,400 lbs.	19,600 lbs.	30,240 lbs.
11				13,350 lbs.	18,200 lbs.	29,650 lbs.
12				12,200 lbs.	16,600 lbs.	28,800 lbs.
13				10,500 lbs.	14,350 lbs.	27,700 lbs.
14				8,950 lbs.	12,200 lbs.	26,300 lbs.

Table 6-4 Maximum load capacity for support columns constructed of No. 1 Common Grade: Southern Pine, North Carolina Pine and Douglas Fir.

Nominal size (in.)	3 × 4	4 × 4	4 × 6	6 × 6	6 × 8	8 × 8
actual size (in.)	2⅝ × 3⅝	3⅝ × 5⅝	3⅝ × 5⅝	5½ × 5½	5½ × 7½	7½ × 7½
area	9.51	13.14	20.39	30.25	41.25	56.25
Column height (ft.)						
4	8,720 lbs.	12,920 lbs.	19,850 lbs.	30,250 lbs.	41,250 lbs.	56,250 lbs.
5	7,430 lbs.	12,400 lbs.	19,200 lbs.	30,050 lbs.	41,000 lbs.	56,250 lbs.
6	7,630 lbs.	11,600 lbs.	17,950 lbs.	29,500 lbs.	40,260 lbs.	56,250 lbs.
6½	4,750 lbs.	10,880 lbs.	16,850 lbs.	29,300 lbs.	39,950 lbs.	56,000 lbs.
7	4,130 lbs.	10,040 lbs.	15,550 lbs.	29,000 lbs.	39,600 lbs.	55,650 lbs.
7½		9,300 lbs.	14,400 lbs.	28,800 lbs.	39,000 lbs.	55,300 lbs.
8		8,350 lbs.	12,950 lbs.	28,150 lbs.	38,300 lbs.	55,000 lbs.
9		6,500 lbs.	10,100 lbs.	26,850 lbs.	36,600 lbs.	54,340 lbs.
10				24,670 lbs.	33,600 lbs.	53,400 lbs.
11				22,280 lbs.	30,380 lbs.	52,100 lbs.
12				19,630 lbs.	26,800 lbs.	50,400 lbs.
13				16,920 lbs.	23,070 lbs.	47,850 lbs.
14				14,360 lbs.	19,580 lbs.	44,700 lbs.

Table 6-5 Maximum safe load-bearing capacities for support columns constructed of steel pipe.

Nominal size (in.)	6	5	4½	4	3½	3	2½	2	1
Length/ft.									
5	72,500	55,900	48,000	41,200	34,800	29,000	21,600	12,200	7,50
6	72,500	55,900	48,000	41,200	34,800	28,600	19,400	10,600	6,00
7	72,500	55,900	48,000	41,200	34,100	26,300	17,300	9,000	5,00
8	72,500	55,900	48,000	40,100	31,700	24,000	25,000	7,400	4,20
9	72,500	55,900	46,400	37,600	29,300	21,700	12,900	6,600	3,50
10	72,500	54,200	43,800	35,100	26,900	19,400	11,400	5,800	2,70
11	72,500	51,500	41,200	32,600	24,500	17,100	10,300	5,000	
12	70,200	48,700	38,500	30,000	22,100	15,200	9,200	4,100	
13	67,300	46,000	35,900	27,500	19,700	14,000	8,100	3,300	
14	64,300	43,200	33,300	25,000	18,000	12,900	7,000		

also contain enough flexibility so that normal activity within the structure will result in noticeable vibration or bending of the member, possibly resulting in ceiling or wall cracks, particularly with plaster walls and ceilings.

Calculating joist size

When sizing joists, one must consider the span, the anticipated live load of the structure, the spacing between centers, and the fiber-bending stress of the material to be used. Once this information is known, a tabular method can be used to determine the correct size joists. The following tables present information concerning the dimensions of joists made from material assumed to have a fiber stress of 1200 pounds per square inch.

Calculating the number of joists required

Calculating how many joists are needed will of course be directly related to the spacing between centers of the joists. For illustration purposes, assume the spacing is 16 inches on center spanning the width of a building 48 feet long and 20 feet wide. The formula equals:

Length of building times 75 percent + 1

$$48 \times 0.75 = 36$$

$$36 \text{ joists} + 1 = 37$$

Table 6-6 Maximum joist spans for uniformly loaded joists of varying sizes.

Live load l\b. sq. ft.	Spacing	2×6	2×8	2×10	2×12	2×14	3×6	3×8	3×10	3×12	3×14
10	12	12'9"	16'9"	21'1"	24'0"		14'7"	19'3"	24'0"		
	16	11'8"	15'4"	19'4"	23'4"	24'0"	13'6"	17'9"	22'2"	24'0"	
	24	10'3"	14'6"	17'3"	20'7"	24'0"	11'11"	15'9"	19'10"	23'9"	24'0"
20	12	11'6"	15'3"	19'2"	23'0"	24'0"	13'3"	17'6"	21'9"	24'0"	
	16	10'5"	13'11"	17'6"	21'1"	'0"	12'0"	16'1"	20'2"	24'0"	
	24	9'2"	12'3"	15'6"	18'7"	21'9"	10'6"	14'2"	17'10"	21'6"	24'0"
30	12	10'8"	14'0"	17'9"	21'4"	24'9"	12'4"	16'4"	20'5"	24'5"	
	16	9'9"	12'11"	16'3"	19'6"	22'9"	11'4"	14'11"	18'9"	22'7"	26'4"
	24	8'6"	11'4"	14'4"	17'3"	20'2"	10'0"	13'2"	16'8"	19'11"	23'4"
40	12	10'0"	13'3"	16'8"	20'1"	23'5"	11'8"	15'4"	19'3"	23'1"	26'11"
	16	9'1"	12'1"	15'3"	18'5"	21'5"	10'8"	14'0"	17'8"	21'3"	24'10"
	24	7'10"	10'4"	13'1"	15'9"	18'5"	9'4"	12'4"	15'7"	18'9"	22'1"
50	12	9'6"	12'7"	15'10"	19'1"	22'4"	11'0"	14'7"	18'4"	22'0"	25'8"
	16	8'7"	11'6"	14'7"	17'6"	20'5"	10'0"	13'4"	16'10"	20'3"	23'8"
	24	7'3"	9'6"	12'1"	14'7"	17'0"	8'10"	11'9"	14'10"	17'10"	20'10"
60	12	9'0"	12'0"	15'2"	18'3"	21'4"	10'6"	14'0"	17'7"	21'1"	24'7"
	26	8'1"	10'10"	13'8"	16'6"	19'3"	9'7"	12'10"	16'1"	19'4"	22'7"
	24	6'8"	8'11"	11'3"	13'7"	15'11"	8'5"	11'3"	14'1"	17'0"	20'0"
70	12	8'7"	11'6"	14'6"	17'6"	20'6"	10'1"	13'5"	16'11"	20'5"	23'9"
	16	7'8"	10'2"	12'10"	15'6"	18'3"	9.3"	12'3"	15'5"	18'7"	21'10"
	24	6'5"	8'5"	10'7"	12'9"	15'0"	8'0"	10'7"	13'4"	16'1"	18'10"

To this figure must be added any additional joists required for doubling purposes, such as under-wall partitions running parallel to the joists and headers around any openings. If, for example, ten areas require joists to be doubled, ten more joists would be added to the product of this formula, resulting in a total number of 47 joists needed. Although most construction is planned to accommodate standard 10-foot, 12-foot, 14-foot, or 16-foot lengths of joists, a waste factor of 5 percent is still advisable, which rounded down would bring the total number of joists needed to 49 ($47 \times 0.05 = 2.35$ [rounded to 2]). Floor joist and ceiling joist sizes are determined in the same manner as are the number of joists required.

Headers

Joists are often doubled or tripled to support the load carried by a header and are then referred to as *trimmers*. Just as a support column will be placed under load when supporting a beam, trimmers will be supporting the load of the header, so the load carried by the header must be determined first. The formula for calculating the amount of weight the header will carry is as follows: half the product of the length of the tail beam, multiplied by the length of the header, multiplied by the total pounds per square foot of floor load. Plugging numbers into the formula to represent the length of the tail beam (24), the length of the header (12), and the total pounds per square foot of floor load (40), the mathematical expression would be as follows:

$$\frac{1}{2}(24 \times 12 \times 40) = 5760 \text{ pounds}$$

This equation delivers the total uniformly distributed load on the header; this figure divided by the number of trimmers represents the concentrated load on each trimmer. In this example, the uniformly distributed load on the header would be 5760 pounds; assuming two trimmers, you'd come up with 2880 pounds as the concentrated load on each trimmer.

This example ties in with a principle related to beams joist and headers: a concentrated load applied at the center of a span produces the same effect as a uniformly distributed load of twice the weight. Using this principle when dealing with situations concerning concentrated loads applied to the center of a span, the previous beam tables in this chapter will provide accurate sizing information by doubling the load and treating it as if it were evenly distributed.

Similarly, there are relationships between a load that is uniformly distributed and concentrated loads that are located at points other than the midpoint of the span. Table 6-7 shows the relationships between uniformly distributed and concentrated loads as they are effected by the position of the concentrated load.

Whatever the weight of the concentrated load, determine the position of that load on the span (¼, ⅓, ½, etc.), find the corresponding multiplier in the table and do the math. For example, for a 2100-pound concentrated load located at ⅓ the span, the multiplier is 1¾ (or 1.75). The corresponding uniformly distributed load would be determined as:

$$2100 \text{ pounds} \times 1.75 = 3675 \text{ pounds}$$

This number can now be used in determining the appropriate size joists from the tables dealing with maximum joist span found earlier in this chapter.

Table 6-7 Correlation factors between concentrated and uniformly distributed loads.

Concentration point of load	Correlation factor (multiplier)
Center of span	Multiply the concentrated load by 2
⅓ of the way along the span	Multiply the concentrated load by 1¾
¼ of the way along the span	Multiply the concentrated load by 1½
⅕ of the way along the span	Multiply the concentrated load by 1¼
⅐ of the way along the span	Multiply the concentrated load by 1
⅛ of the way along the span	Multiply the concentrated load by ⅞
⅒ of the way along the span	Multiply the concentrated load by ¾

Subfloors

Used as a base for finish flooring, subflooring normally consists of plywood sheets or tongue-and-groove boards. The choice of subflooring is largely dependent upon the finish flooring to be installed and the spacing of the joists.

Plywood

Because of its excellent properties as a structural subfloor and substrate, plywood often fulfills two functions: as subflooring and as underlayment. Determining the size (thickness) of the plywood sheets to be installed is directly influenced by the type of wood used and the spacing between joists. For installations where southern pine or West Coast Douglas fir is used, the minimums are as follows:

24-inch joist spacing	Minimum ¾-inch thickness
20-inch joist spacing	Minimum ⅝-inch thickness
16-inch joist spacing	Minimum ½-inch thickness

If ponderosa pine, western hemlock or western white pine are used, the minimums are somewhat different, as follows:

24-inch joist spacing	Minimum ⅞-inch thickness
20-inch joist spacing	Minimum ¾-inch thickness
16-inch joist spacing	Minimum ⅝-inch thickness

Installed with the direction of the grain of the outer plies at right angles to the joists, plywood sheets should be staggered to ensure that the end joints of adjacent sheets do not break over the same joists.

Boards

Tongue and groove or square-edge boards, 8 inches or less in width and ¾ of an inch or greater in thickness, are also used for subflooring needs. This type of subflooring is most often laid in a diagonal pattern to the joists for the greatest versatility. When laid in a diagonal pattern, the finish flooring may be laid parallel with the joists, or perpendicular to the joists. If the boards of the subflooring are laid at right angles to the joists, the finish flooring should always be installed at right angles to the subflooring. In either case, the end joints of the subflooring boards should be made directly over the joists.

Determining amounts needed

Regardless of whether plywood or boards have been specified as the subfloor material, calculating the quantity of subflooring material required begins with determining the area of the space to be covered.

Once the area of the room (length times width) is known, the rough quantity of subflooring boards is also known. Since subflooring boards are normally made of 1-inch lumber, the area of the room is equal to the number of board feet (1-foot long by 1-foot wide by 1-inch thick) required for that room. Additions to this rough quantity must be made to accommodate special features of the room, such as bay windows. Other additions must be made for waste factors. As a rule of thumb, 15 percent should be added if the subfloor is laid at right angles to the joists. If the subfloor is laid diagonally, a 20-percent waste factor should be added to the materials estimate.

Calculating the amount of plywood sheets needed for the project is a simple matter of dividing the area of a sheet a plywood into the total area to be covered. As with board subflooring a normal 15-percent waste factor should be added to the total. If there are projections such as bay windows to account for remember to allow additional plywood for adequate coverage and increase the waste factor to 20 percent.

Review questions

1. A home designer needs to specify a header size needed to carry the roof members over a 9'6" opening for a window design. The roof is simply supported and has a span of 40'0". Assume a 30#/ft^2 live load to accommodate snow and workmen, and a 15#/ft^2 dead load for lumber and roofing materials. Determine the uniform load (in #/ft) the window header will carry.
2. From Table 6-8 choose a member size to carry the load. Simple supported framing members are supported at each end only, no intermediate supports are used.

Table 6-8 Glued laminated floor and roof beams - span data

Span data for glued laminated roof beams
Maximum deflection 1/240 of the span

Beam size (actual)	Beam weight per lin. ft. in pounds	Span in feet 10	12	14	16	18	20	22	24	26	28	30	32
		Pounds per lin. ft. load bearing capacity											
3 × 5¼	3.7	151	85										
3 × 7¼	4.9	362	206	128	84								
3 × 9¼	6.7	566	448	300	199	137	99						
3 × 11¼	8.0	680	566	483	363	252	182	135	102				
4½ × 9¼	9.8	850	673	451	299	207	148	109					
4½ × 11¼	12.0	1,036	860	731	544	378	273	202	153				
3¼ × 13½	10.4	1,100	916	784	685	479	347	258	197	152	120		
3¼ × 15	11.5	1,145	1,015	870	759	650	473	352	267	206	163	128	104
5¼ × 13½	16.7	1,778	1,478	1,266	1,105	773	559	415	316	245	193	154	124
5¼ × 15	18.6	1,976	1,647	1,406	1,229	1,064	771	574	438	342	269	215	174
5¼ × 16½	20.5	2,180	1,810	1,550	1,352	1,155	933	768	586	457	362	290	236
5¼ × 18	22.3	2,378	1,978	1,688	1,478	1,308	1,113	918	766	598	478	382	311

3. Mr. and Mrs. Johns are planning a new home. They want as much window area as possible. The local energy code permits a maximum window area of 17% of the house floor area. The windows John and Joan will use are each 3 ft × 5 ft and the floor area of the house is 1720 square feet. How many windows can they put into their new house?
4. The cost of hardwood flooring is $7.00/square foot and the installation cost is $2.00/square foot.
   a. A family room measures 18 ft × 24 ft. What is the cost for hardwood flooring, including installation?
   b. How many standard-sized sheets of plywood will be needed as subflooring for the family room described?
5. An I-beam 12-feet long weighs 52 pounds. How much does an I-beam of the same width weigh if it is 18 feet long?

7

Math for wall framing

For the most part, this chapter focuses on the elements of conventional platform framing construction. Soleplates, top plates, studs, bracing, headers, trimmers, and fire blocks constitute the supporting structure of a wall.

Wall framing

Bearing walls serve several functions besides holding their own weight:

- Bearing walls support the weight of the structure above the top plate of the bearing wall.
- They provide a secure framework onto which is attached exterior insulation, sheathing and the exterior finish.
- The cavities formed between the studs provide an ideal opportunity to further insulate the structure by installing fiberglass batts within them.
- The interior face of the insulated wall framing now becomes the framework on which drywall and an interior wall covering are normally placed.

Generally, the lumber used in the components of walls is 2-inch by 4-inch stock, with the notable exception of door and window headers (or lintels) within load-bearing walls, which are customarily constructed with 2-inch-by-6-inch lumber. Depending on the length of the window or door opening's span, headers may be single or double members.

Determining the amount of studs needed

To speed up the process of figuring out how much lumber will be required for studs, plates, and caps, use the following procedure. Assuming 16-inch on-center construction for all the walls and partitions, figure out the total linear footage for which lumber is required:

1. Add together the footage of the lengths of all walls and partitions and multiply by ¾ (or 75 percent). For example, 200 feet of walls and partitions × ¾ = 150 feet.
2. Count the number of wall and partition segments and multiply by the length of one stud (8 feet). Assuming there are a total of ten walls and partitions, 10 × 8 feet = 80 feet.
3. Count the number of times an exterior wall is intersected by a partition. Multiply this number by the length of one stud (8 feet). If this situation occurs five times, 5 × 8 feet = 40 feet.
4. Count the number of times one partition intersect another partition. Multiply this number by the length of one stud. If there are four instances where one partition intersects another, 4 × 8 feet = 32 feet.
5. Count the number of openings in all walls and partitions. Double this figure and multiply by the length of one stud. For illustration, if you assume a total of 15 openings, the equation would be 15 × 2 = 30, and 30 × 8 feet = 240 feet.
6. To allow for plates and caps, multiply the answer from #1 (150 feet) by 3, then divide that answer by the length of one stud (8 feet). In this example, the equation would read 150 × 3 = 450, so 450 ÷ 8 feet = 56¼ (or 56.25) feet.
7. Total the answers from each step: 150 + 80 + 40 + 32 + 240 + 56¼ = 598.25 feet.
8. Divide the total from #7 by the length of one stud: 598.25 feet ÷ 8 feet = 74.78 studs.

9. Allowing 5 percent for waste (3.74 studs), round up to a total of 79 studs.

Steel framing

Light-gauge steel introduces an entirely new dimension and potential for error to conducting takeoffs and estimating labor costs. The challenges that a framer faces include having to estimate labor costs for unfamiliar assemblies and deal with materials that are new to the people in the field.

Takeoffs present a situation where call-outs from plans are for materials that estimators are only mildly familiar with. Another situation involves estimators looking at plans that simply do not have details for many openings and assemblies because the engineer did not know how to detail a particular design using metal.

Metal differences

With wood, assembling a window opening takes approximately ten minutes. You receive a 4″ × 10″ Header from your supplier, nail it directly to the King Stud and spread and buck-up the top and bottom Cripple.

When estimating for steel, the obvious difference is using screws instead of nails; cuts involve sawing steel, not wood; and stud sizes larger than 2″ × 4″ C-Studs must be assembled. Calculating labor costs requires accounting for the added time needed to assemble each steel component.

Assembling a window opening with steel instead of wood typically requires:

- assembling the Header by clamping two C-Studs together and applying twelve #10 screws,
- attaching four L90 fasteners to the Headers so that the Header can be attached to the King Stud,
- assembling the top and bottom Cripple with a deep leg track and attaching four screws on the top and bottom,
- attaching the four L90s to the King Stud with twenty-four #10 screws.

This entire assembly takes approximately thirty-five minutes. Most components of the steel home are not as complicated as the window openings and do not require three to four times more labor than for wood. However, the fact that there is a significant amount of additional time required for many components of the home necessitates precisely breaking down every structural element when adding up labor costs.

Learning how to assemble the different components of the house and calculating the labor required for each task is only half the battle. Labor costs that differ from those of wood must be combined with the differences in engineering from one set of plans to another. Currently there is a wide range of acceptable industry interpretations on standards for steel. Two separate sets of plans may call for a window to be assembled differently. Shooting from the hip on such an estimate would be ruinous. Two different sets of plans, designed by two different engineers, can easily have completely different structural requirements. Differences in the plans include everything from the gauge of steel to the number of screws attaching the sheer to the studs. One set of plans may call for twelve screws on both ends of the L90s, while the other calls for only four screws. Estimators not only have to account for the added hardware costs, but also for the fact that it may take the carpenter twice as long to do the assembly.

The American Iron and Steel Institute along with groups of contractors and engineers are working toward standardizing the metal framing industry. Many efficiencies have been gained from experience in the field, and metal framing is gaining acceptance rapidly. Become familiar with the components of metal framing. Line-up dependable suppliers who can help educate and direct your crews. But most of all take the time to properly estimate each phase of a metal framing job.

Partition framing

Partition walls, whether load-bearing or not, include most of the same elements and are, for the most part, constructed in the same manner as are load-bearing exterior walls. Both bearing and non-bearing walls require soleplates, studs, bracing, and necessary headers. However a non-bearing partition wall will require only one top plate (often called a *cap*). Frequently, partition walls will be erected as part of the overall framing at the same time as the exterior walls are constructed. If, however, the floors are not to be installed until after the exterior of the structure is complete, the partition walls can be framed anytime after the flooring is in place. Determining the amount of material necessary for partitions requires exactly the same procedures as used in determining materials needed for bearing walls.

Framing for windows & doors

In the construction of load-bearing partitions and walls, trimmers support the lintels at the ends and are often directly joined to the studs. In all such cases, the lintels should be doubled to make sure the load is properly supported, regardless of the width of the opening. Fabricating a doubled member, by placing one member on top the other to make up a header, requires that the two members be securely attached to each other in order to ensure proper support for the load which they will bear.

Headers intended for use in door or window openings greater than 3 feet wide will require more strength to adequately support the load imposed on them. One way in which this additional support can be obtained is by using material with a larger cross section than that offered by a 2-×-4. Another alternative is to use a truss constructed of 2-×-4 material. A truss of this nature will generally be sufficient to handle the additional load resulting from openings of 3 to 6 feet in width. However, for spans 6 feet wide or greater a load bearing beam should be used to accommodate the weight. The tables and equations previously discussed in Chapter 6 of this book will be useful in calculating the type and size of material to be used for such beams. Table 7-1 provides a range of appropriate header sizes which correspond to various widths of openings. For normal light-frame construction, doubling members made of materials normally used for floor joists is a commonly accepted practice and will offer sufficient support.

To allow for proper clearance of head casings, frames, and flooring; the studs, headers, and sills around wall openings need to correspond to the dimensions recommended by the window and/or door manufacturers. Depending on the type and thickness of the finish flooring to be installed, the bottoms

Table 7-1 Header sizes as related to opening size.

Max. width of opening (in feet)	Corresponding header size (in inches)
3½	2 by 6
5	2 by 8
6½	2 by 10
8	2 by 12

of window and door headers can usually be set at 6 feet 10 inches above the subfloor surface. The rough framing sizes needed for exterior windows and doors will vary depending on the manufacturer, however, the following guidelines will generally be acceptable. For individual double-hung windows, the approximate rough opening width will equal the width of the glass plus 6 inches. The approximate rough opening height will equal the total glass height plus 10 inches. For casement windows, the width of the rough opening will be approximately equal to the total glass width plus 11¼ inches, while the height of the rough opening will be approximately equal to the total glass height plus 6⅜ inches. The approximate width of an exterior door's rough opening will equal the width of the door plus 2½ inches, and the approximate height will equal the height of the door plus 3 inches.

Ceiling joists

Ceiling joists, as is the case with many construction members, serve the structure in more than one capacity. Along with supporting ceiling finishes, ceiling joists often double as the floor joists of the next level while also acting as ties between exterior walls and interior partitions.

Headers will need to be placed at right angles to the ceiling joists in order to carry the tail beams when any opening is cut through the joists, as for a stairway access for example. Depending on requirements, headers used in these circumstances should be supported by double or triple trimmers. In order to prevent a projection below the ceiling line, the trimmers should never be of a depth greater than the depth of the joists themselves. The actual loads to be carried by the headers and trimmers will determine the size of these members. As mentioned in Chapter 6, the tables and equations found in that chapter will simplify the sizing process greatly. As a general guideline, however, the header will support the weight of the floor to the midpoint of the tail beams which rest on it. Mathematically expressed, the equation would be:

$$(\text{Length of tail beam} \times \text{length of header} \times \text{total square foot floor load})\backslash 2$$

Remember, when performing calculations, any functions stated within parenthesis must be completed prior to any other steps in the equation.

Exterior walls

Many different forms of exterior finish materials exist today. Some materials are suited to one climate more than another, while some are more universally suited to their purpose. Certain materials and their application techniques can greatly affect a structure, increasing the buildings resistance to air infiltration and lending additional support to the walls with increased rigidity are two examples. Some of the more common types of exterior sheathing material are discussed here.

Gypsum sheathing

Gypsum sheathing is used as a protective fire-resistant membrane under exterior wall surfacing materials such as wood siding, masonry veneer, stucco, and shingles. It also provides protection against the passage of water and wind and adds structural rigidity to either wood or metal framing systems. The non-combustible core is surfaced with water-repellent paper; in addition it may also have a water-resistant core. Some available boards are covered with fiberglass mat that provides an alkali-resistant surface. This facing also resists wicking, moisture penetration, and de-lamination caused by surface water exposure. The alkali-resistant surface coating means that it is not necessary to apply a primer/sealer with an exterior installation finish system.

Gypsum sheathing is commonly available in 2- and 4-foot widths, 8 feet in length, and ½-inch or ⅝-inch thicknesses and is available with a type-X core.

Gypsum sheathing panels may be applied horizontally, (long edges of the board at right angles to the framing members) or vertically (long edges parallel to the framing). Horizontal drywall application is generally preferred because it offers the following advantages:

- Reduces the lineal footage of joints up to 25 percent.
- The strongest dimension of each board runs across framing members.
- The sheets bridge irregularities in the alignment and the spacing of framing members.
- Each board ties more framing members together than do boards in parallel applications, thus providing better bracing strength.

Applied vertically, each sheet of gypsum will normally cover three stud spaces when the studs are spaced 16 inches on center, and two studs when the spacing is 24 inches on center. The edges of each gypsum sheathing panel should be centered on studs, and only moderate contact should be made between edges of sheets.

For wall applications, if the plate height is 8 feet 1 inch or less, horizontal application will result in fewer joints, easier handling, and less cutting. If the plate height is greater than 8 feet 1 inch or the wall is 1 foot wide or less, vertical or parallel application will prove to be more practical.

Although panels longer than 8 feet are priced slightly higher than the standard-sized panels, it is generally advisable to order the longest panel which is usable; the difference in price is worth it because the panel covers a larger area with less time and reduces the number of joints. (But remember that smaller boards are much easier for one person to handle.) If, for instance, the wall to be sheathed is 21 feet long, it would be better to use one 4-×-12 panel and one 4-×-10 panel rather than three 4-×-8 panels for each horizontal course. Not only would there be a smaller amount of waste, there would also be less labor involved. In such a case the little extra cost of the larger sheets would be justified in the time saved and the neatness of the finished job.

Plywood sheathing

A popular material for sheathing, plywood sheets normally measure 4 × 8 feet but are available in longer sizes. Plywood sheathing is usually installed vertically and effectively eliminates the need for diagonal corner bracing.

Wood sheathing

Nominal 1-inch boards manufactured with a tongue and groove, square-edge, or shiplap pattern are most commonly the material used as wood sheathing. Important characteristics of wood sheathing are its minimal moisture content, which will limit the amount of shrinkage after installation and its workability (in other words, the sheathing should be easy to nail through with little if any splintering or splitting).

Insulating board sheathing

Insulating board sheathing is treated with a weather-proofing agent (usually asphalt) to make it water-resistant. Normally the insulating board sheathing (also referred to as fiberboard) is either impregnated with, or coated one or more times on all sides with, the weather-proofing agent. Because of the insulating fiberboard's water resistance, this type of sheathing is more forgiving of the occasional foul weather encountered while the product is being stored on job sites or even between the time it is applied and the time it is covered with the exterior finish material.

Of the three common types of fiberboard used for sheathing purposes, the least dense is referred to as regular density. Regular density fiberboard sheathing is normally available in sheets of 2 feet by 8 feet, or 4 feet by 9 feet and in thicknesses of ½ inch and 25⁄32 of an inch.

Nail base fiberboard and intermediate density fiberboard are both more dense than regular density fiberboard. Both nail base and intermediate density sheathing materials are commonly manufactured in 4 foot by 8 foot and 4 foot by 9 foot sheets. Due to weight considerations, intermediate density and nail base fiberboard are normally limited to a thickness of ½ inch.

Calculating quantities of sheathing required

Determining the number of sheathing panels required for a job is essentially the same as for figuring the necessary amount of subflooring for a job. Multiplying the height of each wall by the width of each wall to be sheathed will result in the total area of that wall to be covered. Dividing this figure by the area of a single sheet of sheathing will result in the number of sheets of sheathing necessary to do the job. For example, assume a simple one-story structure with four sides. Given that each side of the structure measures 10 feet in height and is 20 feet in width and is to be sheathed with 4 by 8 sheets (regardless of the type of sheathing panel), the calculations would be as follows:

$$10 \text{ feet} \times 20 \text{ feet} = 200 \text{ square feet}$$

$$200 \text{ square feet} \times 4 \text{ walls} = 800 \text{ total square feet}$$

$$\text{One 4-}\times\text{-8 sheet of sheathing} = 32 \text{ square feet}$$

$$\frac{800}{32} = 25$$

The project used in this example would require 25 sheets of sheathing to cover the four sides of the structure. When measuring the area of structure for sheathing, do not subtract the area attributable to window and door openings in the walls. The excess material resulting from the difference between the total gross area and the total net area will provide an automatic waste factor.

Another method which can be adapted for estimating the amount of panels or sheets of sheathing material required for structures which have flat roofs is referred to as the perimeter method. With this method the length of each wall is measured and added together to determine the perimeter of the four walls. Normally used for estimating interior wallboard requirements the perimeter method is based on each story of a structure measuring 8 feet in height and also assumes 4 foot by 8 foot sheets of material are to be used. Table 7-2 provides the corresponding number of sheets of sheathing that would be needed for a variety of different perimeter measurements.

Using the first row of numbers from Table 7-2, the perime-

Table 7-2 The perimeter method of estimating sheathing needs.

Perimeter	Number of 4'-by-8' sheets needed
36 feet	9
40 feet	10
44 feet	11
48 feet	12
52 feet	13
56 feet	14
60 feet	15
64 feet	16
68 feet	17
72 feet	18
92 feet	23

ter is 36 feet, meaning that the length of any four walls added together has equalled 36 feet. Assuming that each story of the structure is 8 feet in height, then 36 feet of perimeter multiplied by 8 feet of height equals 288 square feet. Consequently the number of panels required would be determined by dividing 288 by the area of a 4 by 8 foot panel ($4 \times 8 = 32$ square feet), which equals nine panels, which is the corresponding number of panels presented by the table.

If, as in the previous example, a structure has four walls each measuring 20 feet in length, the perimeter of the building would be 80 feet. In Table 7-2, 80 feet is not listed and falls between two other perimeter measurements. In this situation, use the higher of the two listed perimeter measurements, which in this case would be 92 feet. With a 92-foot perimeter, due to the table being based on story heights of 8 feet, the corresponding number of panels in the table is 23 and would cover 736 square feet of area. In the example, the height of the walls are 10 feet, resulting in 800 square feet of area to be sheathed. By subtracting ½ panel for each door and ¼ to ½ panel for each window and adding a 5-percent waste factor Table 7-2 will result in reasonably accurate estimates.

For structures with more than one story, simply multiply the end result of the preceding calculations by the number of stories to be sheathed.

Thermal Performance

Popular calculation procedures for residential wood-frame construction tends to overestimate the actual thermal performance of many of today's housing designs. This error is due to larger fenestration areas and more exterior wall corners in today's designs than in the past. This leads to the need for a thermal performance indicator to represent the whole wood-frame wall, including thermal shorts created at wall interfaces with other envelope components. For this procedure to gain popular acceptance it must be accurate, yet simple enough to be understood by homebuyers as well as builders, and permit thermal performance comparisons of alternative wall systems to wood-frame walls.

Before getting deeper into this discussion, it is necessary to become familiar with some relatively new thermal terminology.

- *Center-of-Cavity R-value.* R-value at a point in the cross-section of the wall containing the most insulation.
- *Clear-wall R-value.* R-value of the wall area, as seen in Fig. 7-1, containing only insulation and necessary framing materials. In other words, a clear section with no fenestration, corners, or connections between other envelope elements such as roofs, foundations, and other walls.

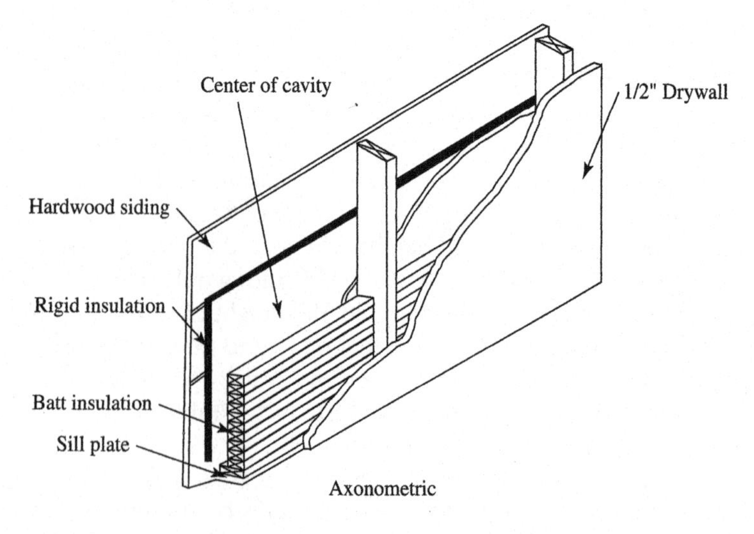

7-1 *Clear wall.*

- *Interface details.* A set of common structural connections between the exterior wall and other envelope components. Examples are wall/wall (corners, Fig. 7-2), wall/roof (Fig. 7-3), wall/floor, window header, window sill, door jam, door header, and window jamb, which make up a representative residential whole-wall elevation.

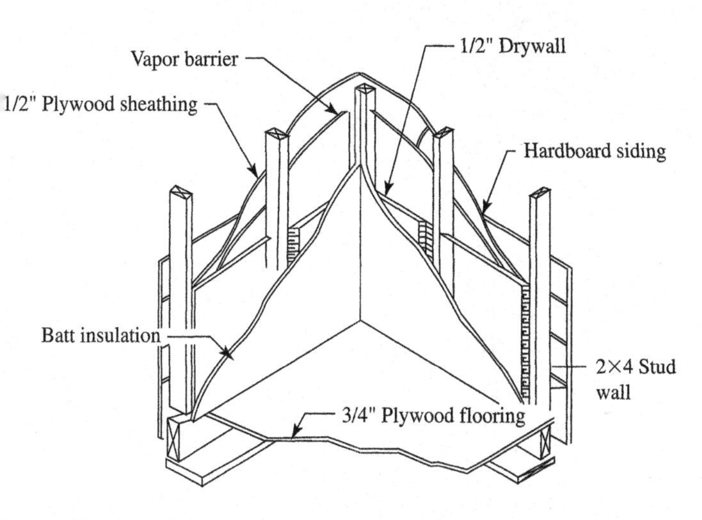

7-2 *Structural connections wall/wall, corner.*

- *Whole-wall R-value.* R-value for the whole opaque wall including the thermal performance of not only the "clear-wall" area, but also all typical envelope interface details including wall/wall (corners), wall/roof, wall/floor (Fig. 7-4), wall/door, and wall/window (Fig. 7-5) connections.

Currently, the percentage reduction of clear wall area R-value to that estimated at the center-of-cavity (thermal evaluation of wood-frame wall systems) is handled by conducting a simple parallel-path calculation for the cavity and stud area. The resulting whole-wall thermal transmittance is compared to the desired value prescribed by either an enforced building energy code, volunteer home energy rating program, or standard. Sometimes, only the center-of-cavity insulation material R-value is used for comparison to alternatives.

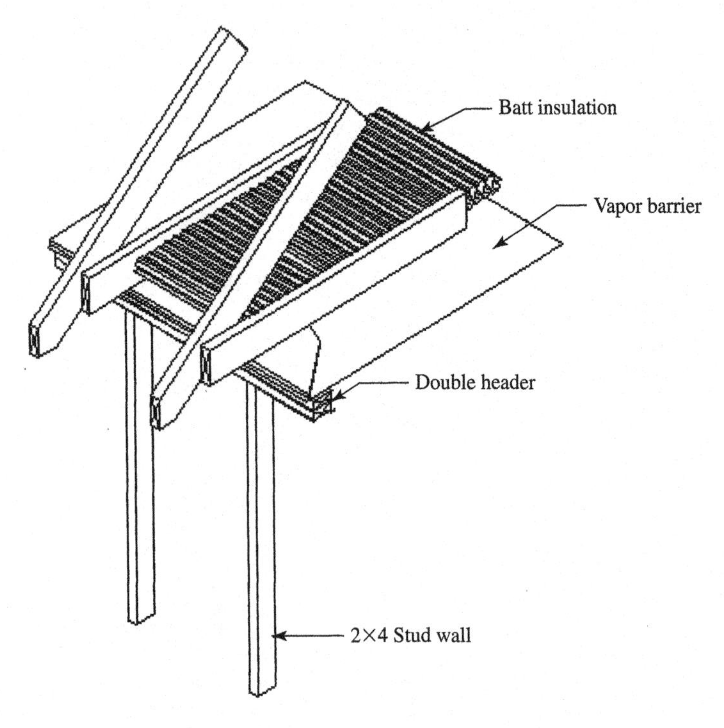

7-3 Structural connections wall/roof.

With today's residential buildings increasingly constructed with materials such as metal, stress skin-insulated core panels, and novel composites, a more accurate rating is necessary. Opaque envelopes can no longer be compared by frequently misleading "center-of-cavity" insulation material or clear wall R-values. The development of more accurate, consumer-understandable wall labels will spur greater market acceptance of energy-efficient envelope systems.

The benefit of advanced systems with only a few thermal shorts compared to conventional wood-frame systems will be clearly discernible by comparing whole-wall thermal performance ratings. The large market share currently held by dimensional wood-frame systems, in part, reflects the misleading and inflated thermal performance ratings currently assigned them. The effect of extensive thermal shorts on performance is

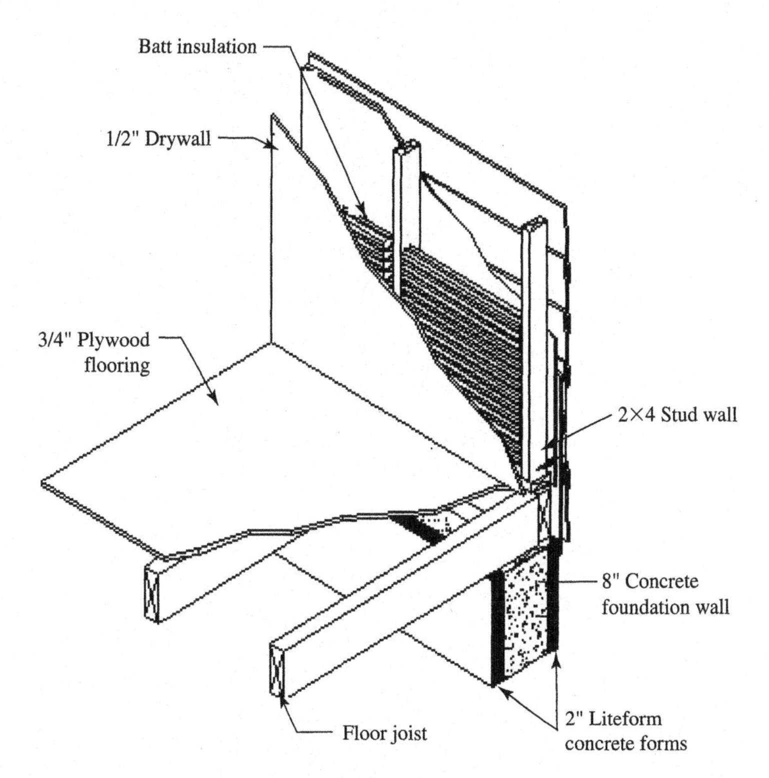

7-4 *Structural connections wall/floor.*

not accurately reflected in commonly used simplified energy calculations that are the current basis for consumer wall thermal comparisons.

Consequently, the market place does not currently account for the thermal shorts that exist in building walls. This results in not realizing the full energy cost savings anticipated by complying with energy code formulas and standards or even meeting the requirements of home energy rating systems. In addition, several building trends suggest that unless more consideration is given to the whole-wall thermal performance, even more energy-saving opportunities will unintentionally be lost.

With the improvement in window efficiency, the potential exists for residential structures to have more windows. When

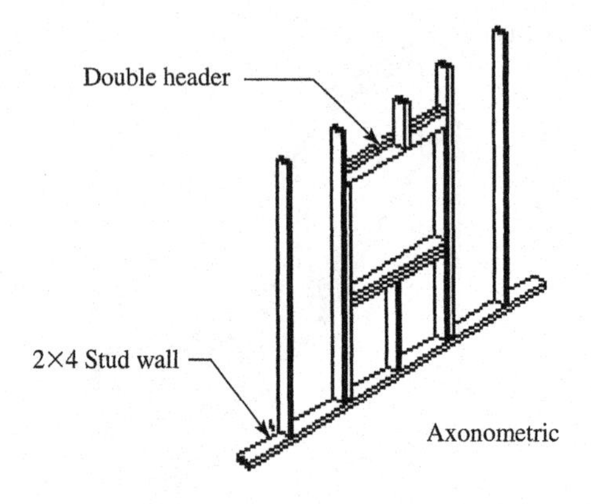

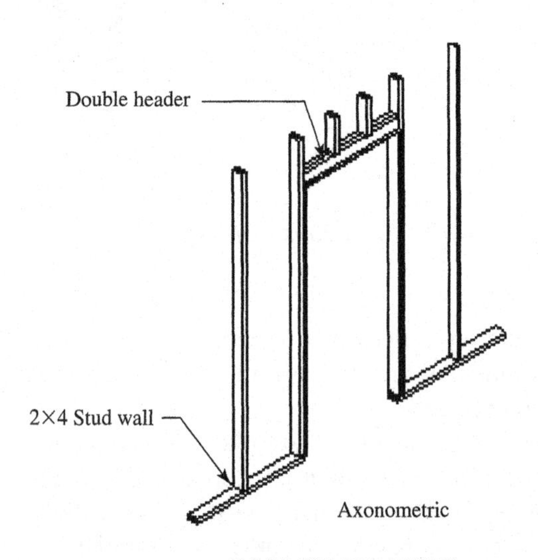

7-5 *Structural connections wall/window.*

more windows are installed in a building, the additional framing that is needed produces a higher overall thermal transmittance of the opaque wall. With metal-frame construction gaining popularity in residential construction, the thermal shorts potentially resulting from the relatively higher thermal conductivity of metal compared to wood can mean much more severe heat loss than is accounted for by traditional, simplified calculations.

The effects of interface details

Interface details are needed to properly baseline the thermal performance of common residential wood-framing systems and to more comprehensively evaluate alternatives. Their inclusion also creates incentives for alternative wall system manufacturers to focus on the whole wall, including the critical connections to other parts of the building, not just the "clear wall." The consequences of poorly selected connections between envelope components are severe. Taking into account the interface details can have an impact on as much as 50% of the overall wall area. For some conventional wall systems, the whole-wall R-value can be as much as 40% less than what is measured for the clear wall section.

With the increasing use of alternatives to dimensional lumber-based systems (such as metal-frame and masonry systems for residential construction), this procedure highlights the importance of using interface details that minimize thermal shorts. Local heat loss through some wall interface details may be double that estimated by simplified design calculation procedures that focus only on the clear wall. Poor interface details also may cause excessive condensation leading to stains and dust markings on the interior finish, which reveal envelope thermal shorts in an unsightly manner. This moist surface area can encourage the propagation of molds and mildews, which can lead to poor indoor air quality.

It has been demonstrated that the whole-wall R-value of residential structures can be determined using a computer. However, this approach requires a level of expertise in three-dimensional, finite-difference heat transfer modeling that is beyond what normally is available in residential building design and construction offices. Therefore, the preferred approach for making this procedure available is a user-friendly interface to a three-dimensional computer model that determines whole-wall R-value for residential buildings. Many software packages that deal with these topics are available for sale. On-line programs sponsored by manufacturers, trade organizations, and educational and governmental institutes can be found on the Internet for use as an interface in these areas. These interfaces will allow users to define the building envelope in terms familiar to the industry rather than in more complex, three-dimensional analytical terms.

Database retrieval tools build upon specific experimental hot-box results, allowing easy modification for particular details and computation of the whole-wall rating for the specific system. The user of this program will see the effect of interface detail improvements and be able to use them in envelope system design and cost optimization.

Performance elements should involve the following:

- Testing full-scale walls under steady-state and dynamic hot-box conditions.
- Three-dimensional, finite-difference computer modeling.
- Thermal analysis of alternative interface details. Hot-box wall tests are used to validate and calibrate two- and three-dimensional computer simulations.
- A steady-state, whole-wall R-value derived for each system.
- To account for thermal mass impacts, if any, customized tables and figures should be generated to reflect dynamic thermal mass benefits compared to low-mass systems.

Within a program, a calculation procedure and ASTM C236 or ASTM C976 test are proposed as a starting point for a consensus methodology for estimating whole-wall R-value, independent of construction type. A clear-wall section, 8 ft by 8 ft (2.4 m × 2.4 m), is tested in a guarded hot box. Experimental results are compared with two- and, if needed, three-dimensional heat conduction model predictions, based on finite-difference methods. The comparison will lead to a calibrated model. This procedure can be performed on any type of clear-wall assembly: metal, masonry, wood, etc.

After the model of the test wall is calibrated, simulations are made of the "clear wall" area with insulation and structural elements and eight-wall interface details, which make up a representative residential whole-wall elevation. These details are as follows:

- Corner,
- Wall/roof,
- Wall/foundation,
- Window header,
- Window sill,
- Door jamb,
- Door header, and
- Window jamb.

Combining results from these detailed computer simulations into a single whole-wall, steady-state R-value estimation, and comparing them with simplified calculation procedures and a result from other wall systems, is the next step. A reference wall elevation must be adapted to weigh the impacts of each interface detail.

Whenever the whole-wall R-value is to be determined, all details commonly used and recommended (outside corner, wall/floor, wall/flat ceiling, wall/cathedral ceiling, doorjamb, window jamb, windowsill, and door header) must be available. The detail descriptions should include drawings, with all physical dimensions, and thermal property data for all material components contained in the details. If the thermal conductivity for one or more critical material components is not available, it may be desirable to measure the individual material conductivity. This may be particularly important if the clear wall hot-box data do not agree with the computer-model predictions. The whole-wall R-value is a better criteria than the clear-wall R-value and much better than the center-of-cavity R-value methods used to compare most types of wall systems. The value includes the effect of the wall interface details used to connect the wall to other walls, windows, doors, ceilings, and foundations.

Exterior paint & stains

Among the various exterior wall finish materials available today, fewer and fewer require periodic painting in order to maintain their aesthetic or structural characteristics. The development and popularity of aluminum siding and vinyl siding with non-chalking, non-fading, long-lasting finishes have done much to replace wood clapboards as an exterior finish for walls. Brick and other natural masonry materials have always been popular exterior finishes, but the ever-increasing costs of the materials and the labor associated with installing finishes of this type is limiting the number of projects that specify them. Among substitutes for brick and stone, such as cement block walls faced with brick veneer, is a relatively new product referred to as external insulating finishing systems (EIFS). A human-made material, EIFS is applied to the exterior surface and then sculpted to imitate brick, stone or stucco finishes. Here again, EIFS is not a product requiring paint to complete its appearance.

Despite the growing popularity of these low-maintenance exterior finishes, wood is still a very commonly used material for siding, shingles, shakes, and trim work. No matter how good the quality of the materials used or the quality of the initial paint job, sooner or later exterior wood finishes will need refinishing.

Manufacturers of paint for exterior surfaces (house paint) have developed formulations designed to eliminate some of the most common problems with exterior finishes, such as chalking and mildew growth. The two major categories of house paint are oil-base and latex. Oil-base paints, with linseed oil and either turpentine or mineral spirits as a thinner, have customarily been used as exterior house paint. However, that is now changing with advancements in latex-base paints becoming more durable. As well as being easier to apply, the water-base latex paints have excellent color retention and take considerably less time to dry. This shorter drying time allows multiple coats to be applied within a 24-hour period, as opposed to oil-base paints which usually require 48 hours or more to completely dry.

Determining paint or stain quantities

On flat surfaces, calculating the amount of paint or stain needed again involves determining the area of the surface to be covered. In the case of sidewalls for example, the height of the wall would again be multiplied by the width of the wall. However, for walls which include a gable end, the calculation must be done in two parts. The first part, as described earlier, includes the section of the wall from the bottom of the roof line to the foundation. The second segment of the calculation measures the height of the gable beginning at the bottom of the roof line and extending to the peak. To adjust for the narrowing of the gable area caused by the pitch of the roof, multiply this height by half of the width of the wall, then add this figure to the area of the wall below the roof line to arrive at the total square footage of the wall to be covered.

Overhang should be figured in two separate segments, the soffit (the portion of the overhand which extends horizontally from the roof line) and the facia (the board behind the gutters or connecting the sidewall to the soffit). Each area should be measured individually, the height and width multiplied, and the results added to the total area to be covered.

Additional portions of structures to be painted or stained, such as dormers, should have their paintable areas individually measured, being careful not to neglect any soffit and facia which may exist around the dormer. As a rule of thumb, however, each dormer will usually account for 100 square feet of coverage area. Unless a wall contains an extra large opening, such as one or more picture or bay windows (which take up most of the wall space), do not subtract for openings; any additional paint or stain will be claimed by extra absorbent material, touch-ups or trim work.

Every gallon of paint or stain includes on its label the average area that it can cover, and these figures are arrived at by the manufacturer through extensive practical testing. Once the total square footage to be painted or stained is known, divide this figure by the coverage figure supplied by the manufacturer to determine how many gallons of product will be required to apply one coat of paint or stain.

Remember, unless the existing finish is in excellent condition, two or more coats will be necessary. For new wood surfaces, one primer coat and two finish coats are usual, and two coats are normal for most project requirements to ensure adequate coverage with no "bleed-through" of the color being covered. After the initial coat of paint or stain, each successive coat will require less product for complete coverage. Table 7-3 provides guideline coverage figures for successive coats of various house paints and stains.

Table 7-3 can be used to adjust the rough figure for the amount of paint or stain required for second or third coats. However, doubling the amount needed for the first coat for

Table 7-3 Average coverage areas per gallon for successive coats.

Type of product applied to type of surface	First coat	Second coat	Third coat
Exterior house paint: Wood siding	420 sq. ft.	520 sq. ft.	620 sq. ft.
Exterior house paint: Shingle siding	342	423	
Shingle stain: Shingle siding	150	225	
Exterior oil/latex paint: Medium texture stucco	153	360	360
Cement water paint: Medium texture stucco	99	135	
Exterior trim paint: Trim	850	900	972

each additional coat will provide sufficient coverage. Any excess product can be used for touch-ups or will be consumed by hard-to-cover areas of the project.

The quantity of paint or stain required for window and door frames, window sash, sill, parting strip, jamb, and casings is almost impossible to figure accurately. Depending on the number and size of individual windows and doors in the structure and the type and quality of stain or paint used, it is better to order too much rather than not enough, to ensure proper color matches—especially if custom-mixed paint colors are being used.

Review questions

1. A site in Salt Lake City needs to buy a truckload of lumber. The requirement is for two different sizes of dimensional lumber in various lengths. The lumber is shipped from Coeur d'Alene, Idaho. Your problem is to calculate the cost of the shipment.

   The supplier works in quantities of 1000 board feet. The mill is asking $380.00 per thousand board feet for the 2 × 4s and $440.00 per thousand for the 2 × 10s. The broker's profit is determined at 4% of the delivered cost.

A standard semi can haul 48,000 pounds of lumber. The lumber ordered weighs 2000 pounds per thousand board feet. Thus, the customer is in effect asking for approximately 24,000 board feet. The customer wants the order divided as 50% 2 × 10s and 50% 2 × 4s, or 12,000 board feet of each size.

The truck needs $1.25 per loaded mile to haul the load. It is 650 miles from Coeur d'Alene, Idaho to Salt Lake City.

The customer wants the following lengths:

2 × 4:	200/8′	200/10′	200/12′	400/14′	400/16′
2 × 10:	80/8′	80/10′	160/14′	160/16′	

Figuring the freight

a. Calculate the shipment composition in board feet for question 1.

b. Calculate a total price for problem question 1.

2. A customer would like a bonus room to be added to an existing home. The builder needs to know the square footage of the new room to figure out how much plywood to buy for the walls. The new room's dimensions are 26' x 22' with 8' ceiling.
3. Using the diagram in Fig. 7-1, calculate the height (ht) of wall A at point (pt) L.

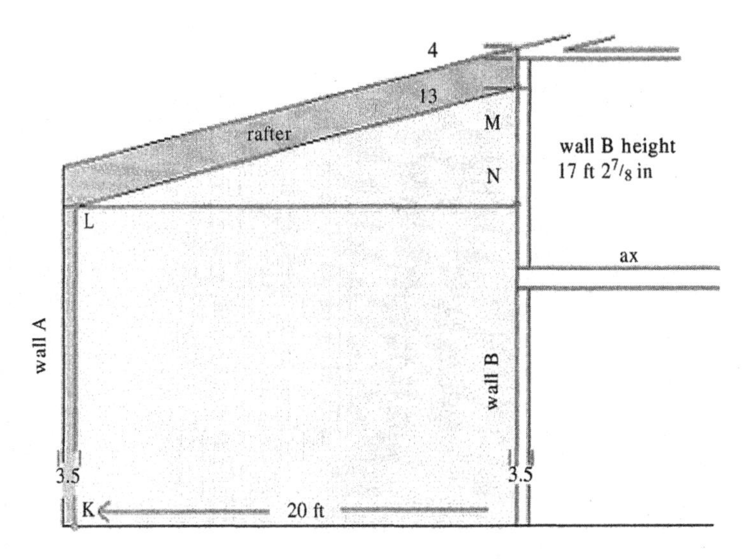

8

Roofing math

The roof of a structure is intended to provide one principal function: sheltering the structure's interior, inhabitants, and their possessions from the elements. Wind, rain, snow, heat, cold, and hail are all forces against which the roof must stand. The roof's function must be given careful consideration when planning its design; it must be properly sloped in order to shed water, snow, and ice, be strong enough to bear up under the extra weight of snow and ice loads, and be constructed to withstand repeated assaults from high winds throughout the years. Shed roofs, hip roofs, gable roofs and valley roofs are only some of the most common forms of roofing construction used.

Determining roof rise & span

The span of a roof is the distance, measured horizontally, from one outside plate (rafter seat) to the outside plate directly opposite. As seen in Fig. 8-1, the span is always measured perpendicularly to the direction of the ridge line of the structure.

The total rise of a roof is the distance, measured vertically (Fig. 8-2), from one outside plate to the center of the ridge line. If a straight line is drawn to connect the ridge line to the plane extending from the outside plate, the horizontal distance of that line (Fig. 8-3) is the total run of the roof. As a general rule, the total run of the roof will be equal to half the span.

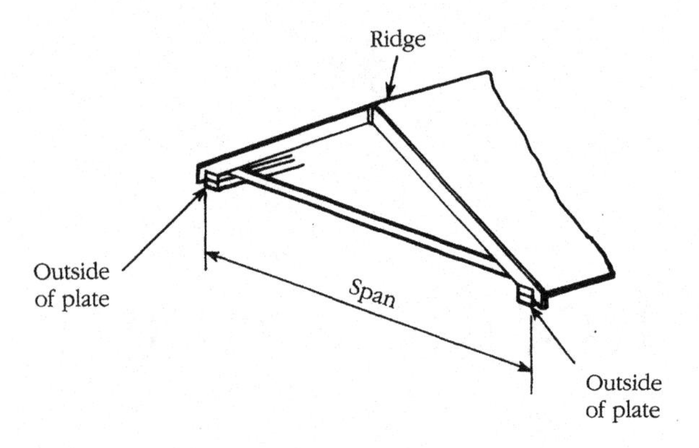

8-1 *The.roof span as shown on a residential building, extends from outside plate to outside plate.*

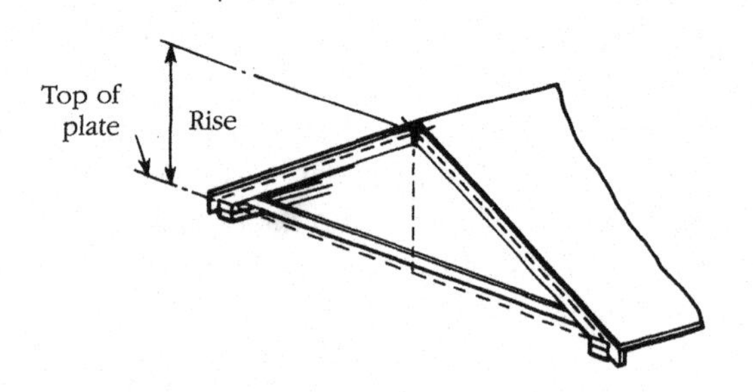

8-2 *Total rise is the vertical distance from eave edge to ridge line.*

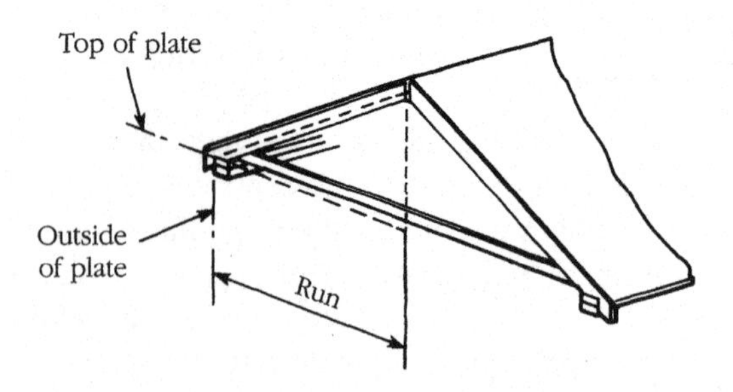

8-3 *Total run is the horizontal distance ridge line to eave edge.*

The number of inches a roof rises per foot of run reveals the slope of the roof and is usually expressed as a ratio. In Fig. 8-4, the roof rises 8 inches for every 12 inches (each foot) of run and would be expressed as 8 : 12 (read "8 to 12"). If the incline of the roof rose 6 inches for each foot of run, the slope would be 6 : 12 and would be a more less severe angle of incline. On the other hand, if the rise of the roof were 10 inches for each foot of run, the slope would be expressed as 10 : 12 and would be quite a steep angle.

If the total span of the same roof is 24 feet with the total rise of 8 feet, the pitch of the roof would be 1/3 (8/24). If the total span of the roof were 32 feet instead of 24 feet with the same total rise of 8 feet, the pitch of the roof would now be 1/4 (8/32).

Rafters

A roof's framework is made up of a number of rafters, performing essentially the same structural tasks as studs perform for walls and joists perform for floors. Rafters are the members connecting the ridge line to the plate, a distance known as the *bridge measure*. Consequently, the length of a common rafter is normally the same as the total run of the roof; and because the run of a roof is usually half the span, the length of a rafter will normally equal half the span of a roof.

As is the case with other structural members, the size of any given rafter will depend on several components, one being the length that member is meant to span (i.e., the distance between the ridge line and the plate). Another is the spacing between each rafter; obviously, rafters of equal dimensions spaced 16 inches on center will support more weight than the same rafters spaced 24 inches on center. And of course, the amount of load the rafters will be expected to carry will play a vital role in determining the dimensions of the rafters.

Common roof layouts

Gable roofs, which feature a high point (ridge) at or near the center of the structure, extending from one end wall to the other, are the most common of the different forms of steep-slope roof styles. The roof slopes downward from the ridge in

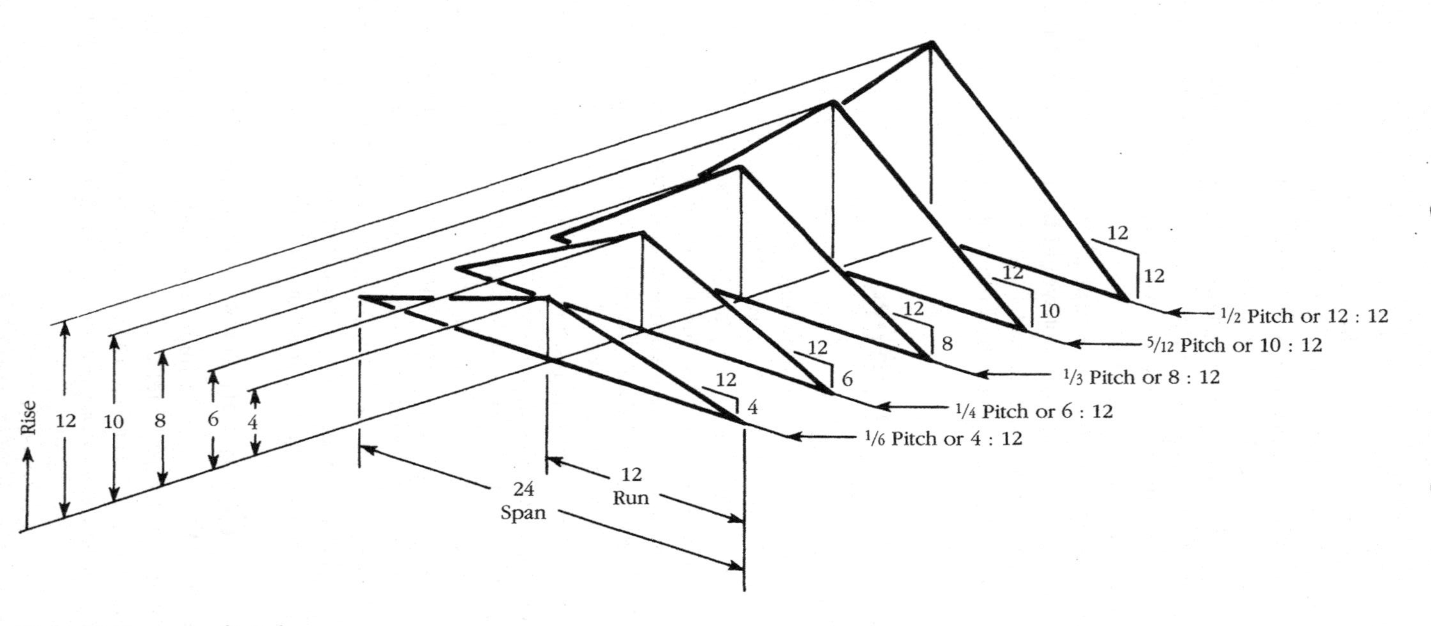

8-4 *Various roof pitches.*

both directions, as seen in Fig. 8-5. The gable takes its name from the triangular section of the end wall between the rafter plate and the roof ridge.

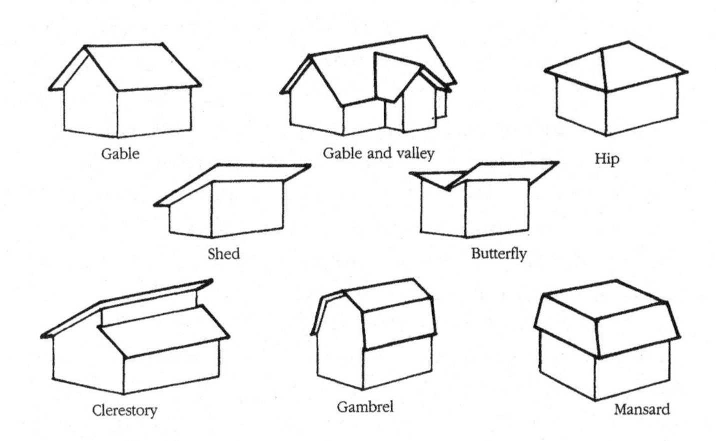

8-5 *Various roof styles.*

A variation of the gable roof is the *gable and valley roof,* which is a combination of two gable roofs each slanting in a different direction coming together to form a valley at the intersection of the two roofs. Although more complicated than other types of roofs to construct, the gable and valley roof is popular in residential building and becoming more popular for small commercial structures.

The dimensions of one side of the roof extending from the plate to the ridge is usually equal to the dimensions, including the slope of the roof section extending from the ridge to the opposite plate.

Hip roofs also have a ridge, but the ridge does not extend from one end of the roof to the other as the gable roof's ridge does. The lower edge of the roof (*eave*) is at a constant height and the roof slopes downward to the eaves on all sides. The point where two roof surfaces meet at an outside corner is called a *hip.* The junction where two roof surfaces meet at an inside corner is referred to as a *valley.*

Shed roofs slope in only one direction and can be visualized by imagining half a gable roof with elongated overhangs placed over the structure. There is no ridge line on shed roofs, and the outside bearing walls supporting the rafters are of different heights, which accounts for the slope of the roof.

Several variations of the shed roof exist. The *butterfly* roof is a configuration of two shed roofs sloping toward a low point located over the center of the structure. Another variation of the shed roof involves two shed roofs rising from opposite eaves but not meeting at a ridge line. Instead, one shed roof meets a vertical plane extending down from the high point of the other shed roof. The wall formed in the space between the two shed roofs is referred to as a *clerestory*. This clerestory is frequently constructed with a series of windows to allow natural light to descend into the structure from the roof.

The *gambrel roof* is often referred to as a *barn roof*, since this style is frequently used in that capacity. It has two sets of slopes: one set gentle, the other set more radical.

Starting at the ridge line, a *gambrel roof* has a gentle slope descending on either side of the ridge to about the midway point between ridge and eave. At this point, the roof takes on a much steeper slope continuing down to the eave edge.

A *mansard roof* is like a hybrid hip roof and is used for commercial buildings, schools and multi-unit housing, such as apartment complexes. As much as 40 percent of the structure can be roof area with this design. With a much shorter ridge line than a hip roof, the mansard roof drops in two distinct slopes to eaves that are the same height all the way around the structure.

Using the steel square's rafter table

Once again, the steel square is proven to be a handy and very versatile tool. Located on the face of the body of the steel square is the rafter table, used to determine the lengths and angles at which common, hip, jack, and valley rafters must be cut for proper distance between and placement to both the ridge and the plate. This table enables the quick and easy determination of

- the line length of any hip or valley rafter per foot of run.
- the line length of any pitch (rafter per foot of run).

- the comparative difference between jack rafters which are spaced 16 inches on center or 24 inches on center.
- side cuts for jack, hip or valley rafters.

The enlarged view of the first line of a rafter table appears in Fig. 8-6. The inches marked on this scale represent the number of inches of rise per foot of run, while the numbers below the graduation marks denoting inches in this row represent the length (given in inches) per foot of run for common rafters. For example, in Fig. 8-6, the number 8 in the top row (8 inches) represents a roof with 8 inches of rise for every 12 inches of run. The number beneath the 8 (14.42 inches) is the length, per foot of run, of the rafters, which would be required for a roof with a slope of 8 : 12.

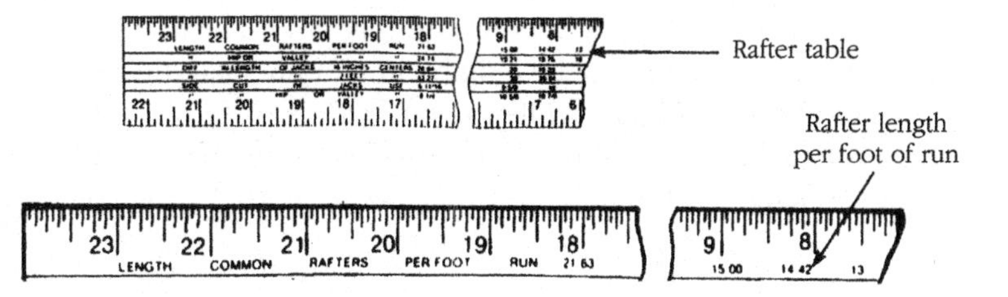

8-6 Segment of a steel square's rafter table.

To determine the length of common rafters for the roof of any structure to have a slope of 8 : 12, use the rafter table on the steel square in combination with the span of the building. If, for example, the span of the building is 34 feet, follow the following steps:

1. Divide the span by 2, to derive the total run of the roof: 34 feet / 2 = 17 feet. (Half the span equals the run, so the run equals 17 feet.)
2. On the rafter table of the steel square, locate the figure representing the required rafter length per foot of run (14.42 inches) as previously found above.
3. Multiply the required rafter length per foot of run (14.42 inches) by the number of feet of the run (17). 14.42 inches × 17 = 245.14 inches
4. Reduce the product of this equation from inches to feet by dividing it by 12. 245.14 inches / 12 = 20.42 feet

By substituting a common fraction in place of a decimal fraction (obtained from Appendix C) the quotient (answer) from Step #4 becomes 20$\frac{5}{12}$ feet, which for standard construction becomes 20½ feet. So 20½ feet would be the line length (actual length) of the rafter in this example. To this length would be added any overhang specified for the roof. That is, with a 20½-foot line length and a 12-inch overhang, the total length of the rafter would be 21½ feet. The length of any overhangs, extending away from the building from the plate line, called for on the plans must always be added to the length of the rafters.

Deriving roof layouts mathematically

Determining the length of roofing rafters is, in fact, calculating the length of the triangle's hypotenuse (Fig. 8-7). The triangle's horizontal base is equal to the length of the horizontal run (half the span of the building), while the other known leg of the triangle is equal to the total rise. The third leg of the triangle, the unknown length of the rafter, is the hypotenuse of the triangle.

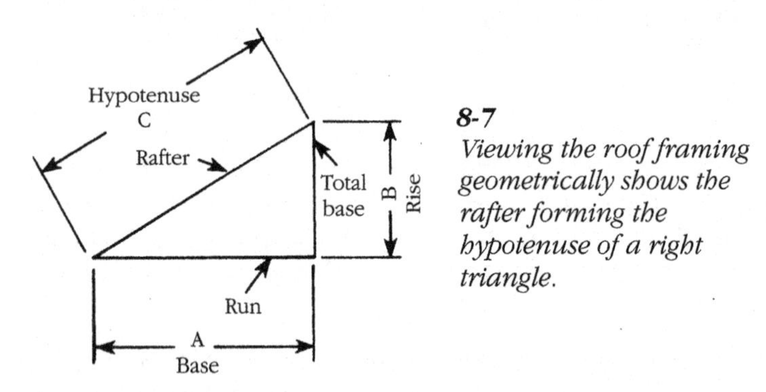

8-7 *Viewing the roof framing geometrically shows the rafter forming the hypotenuse of a right triangle.*

In Fig. 8-8, the base of the triangle (side A) and the vertical leg of the triangle (side B) form a 90-degree angle at their juncture. Because of this 90-degree angle, the resulting triangle which will be formed once the hypotenuse is calculated will be a right triangle.

Now that the unknown length being sought (the length of a rafter) is known to be part of a right triangle, established algebraic theory and formulas can be used to figure out the answer. (Don't get rattled, it's really pretty simple.)

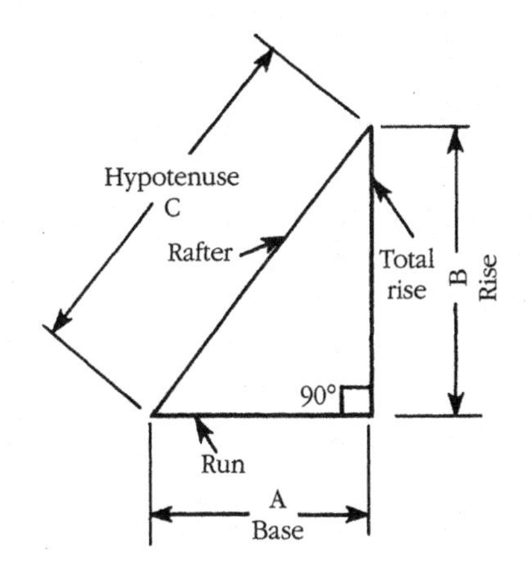

8-8
The two shorter legs of the triangle formed by the roofing members create a right angle.

The Pythagorean Theorem states that, in a right triangle, if A, B, and C are the lengths of the sides, where C is the length of the hypotenuse, then $A^2 + B^2 = C^2$ (A squared plus B squared equals C squared). Remember, any number squared means multiplying that number by itself; for example, 8^2 refers to 8×8, which equals 64. Using random numbers to represent the total run and total rise as in previous examples, and using the Pythagorean Theorem to find the required rafter length, the equation would appear as follows:

1. $A^2 + B^2 = C^2$, where $A^2 = 17$ feet (half the span = run), $B^2 = 22$ feet (rise), and $C^2 =$ the unknown length of the hypotenuse (rafter).
2. Plug the appropriate numbers into the formula $A^2 + B^2 = C^2$: it becomes $17^2 + 22^2 = C^2$, where $17^2 = 289$ and $22^2 = 484$.
3. Plug the appropriate numbers into the formula and the new equation appears as: $289 + 484 = 773$. Remember, all these numbers have been arrived at by squaring another number. A number squared to achieve another number is the *square root* of the new number. (For example, the square root of 289 is 17, since $17 \times 17 = 289$.) To arrive at the correct figure for the hypotenuse of the triangle the square root of the answer derived for C^2, (773) must be calculated.
4. The square root of 773 is 27.80 or converting the decimal into a common fraction by using Appendix C the length of the hypotenuse (rafter) would be $27\frac{19}{24}$ feet.

The Pythagorean Theorem can be used for most applications (such as gable roofs); however, hip roofs are slightly different. Hip rafters and common rafters do have the same unit of rise and both serve as the hypotenuse in a right triangle. However, the unit of run of a hip rafter differs in that it is the hypotenuse of a right triangle with the shorter sides equal to the unit of run of a common rafter. To determine the unit of run of a common rafter, remember that the unit of run for a common rafter equals 12 (12 inches), so the unit of run for a hip rafter would be derived by the following process:

1. Add the product of 12 squared to the product of 12 squared: $(12^2 + 12^2) = 288$.
2. Calculate the square root of the final product from Step #1. The square root of 288 = 16.97.

So the unit of run for a hip rafter is 16.97.

When referring back to the Pythagorean Theorem, with a unit of run equal to 16.97 and a unit of rise of 6, determining the unit length of the hip rafter could be stated as the square root of the quantity 16.97 squared plus 6 squared. The term "the quantity" simply means that parenthesis are placed around that part of the expression and the calculations within those parenthesis are done first. Consequently the procedure would be:

1. $\sqrt{(16.97^2 + 6^2)}$
2. $287.98 + 36 = 323.98$
3. The square root of 323.98 = 18

Simply stated, the results of the calculations mean that for every 16.97 units of run, the hip rafter will have 18 units of rise.

If the total run of the rafter is to be 20 feet, for example, then the length of the rafter is determined by the ratio $16.97 : 18 :: 20 : X$, with X representing the length of the rafter. This proportion can be reduced to the equation $16.97X = 20 \times 18$, which reduces to $16.97X = 360$. Dividing 360 by 16.97 results in the equation $X = 21.21$ feet. So the length of the rafter would equal 21.21 feet.

Since the total run of a common rafter is so easily figured out (half the span), this value can be used to quickly calculate the length of an equal-pitch hip roof rafter. Multiplying the bridge measure by the number of feet in the total run of a common rafter will yield the length of the hip rafter of an equal-pitch hip roof. In the previous example, the span of the structure would be 28²⁄₇ feet, making the total run of a common rafter 14¹⁄₇ feet. Multiplying 14¹⁄₇ feet by 18 (the bridge measure) gives

$$(14\tfrac{1}{7})\ 14.14 \times 18 = 254.52 \text{ inches}$$

$$\frac{254.52 \text{ inches}}{12} = 21.21 \text{ feet}$$

When one converts from decimals to common fractions or vice versa, more than one number will fall into an acceptable range for the conversion number, and any slight differences beyond two decimal places are minor enough to be ignored. For example, in referring to 14.14 feet as 14¹⁄₇ feet, ¹⁄₇ decimally is actually 0.14285714. This means that the common fraction of ¹⁄₇ is 0.00285714 larger than the decimal equivalent of 0.14, and the difference amounts to less than 29 one-thousandths of a foot—hardly anything to be concerned with.

Using the steel square tables

The enlarged second line of the rafter tables on the steel square are shown in Fig. 8-9. The figures on this line of the table are provided to simplify the calculation of hip rafters for hip roofs of equal-pitch. This table can be used similarly to the common rafter table, explained earlier in this chapter.

Note that, in discussions of hip roofs, reference is always made to hip roofs of equal-pitch. Hip roofs are almost always

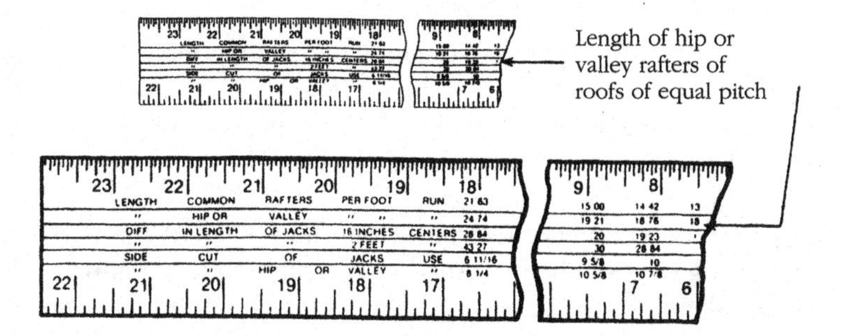

8-9 Enlarged view of the second line of the rafter table on a steel square.

of equal-pitch, but this qualifying statement must still be made in order to ensure the accuracy of the math. Although other various-pitched hip roofs do exist, they are so uncommon so as not to be of concern.

Valley rafter layout

As with hip roofs, most roofs with valley rafters are equal-pitch roofs; unequal-pitch roofs with valley rafters do exist but are scarce. In this book, all valley rafter roofs will be presumed to be of equal pitch.

Valley rafters are necessary when the gable style roof of a dormer, an addition, or a structure wing is tied into the gable roof of the main structure. A valley rafter, as the name implies, lies in the valley formed by the intersection of the main gable roof and any of the additional gable roofs mentioned.

Because all of the gable roofs are assumed to be equal-pitched, the unit of run and unit of rise of a common rafter from the main roof will be equal to the unit of run and unit of rise of a common rafter from any other roof that is tied into the main roof. As seen in Fig. 8-10, valley rafters converge on the ridge of the main roof at a 45-degree angle.

Intersecting roofs of equal span

Under certain conditions, the method of calculating the length of a valley rafter is the same as that for calculating the length of a hip rafter for an equal-pitch hip roof (total run × bridge measure = line length):

- The roofs involved are of equal pitch (as mentioned earlier).
- As seen in Fig. 8-10, the spans of the roofs involved are of equal measure.

These two conditions result in the ridge lines of the two roofs being of equal height. Another contributing factor allowing calculation of the valley rafters length in the same manner as that of hip rafters is that the total run of either of the valley rafters forms the hypotenuse of a right triangle, with the other two sides being individually equal to the run of a common

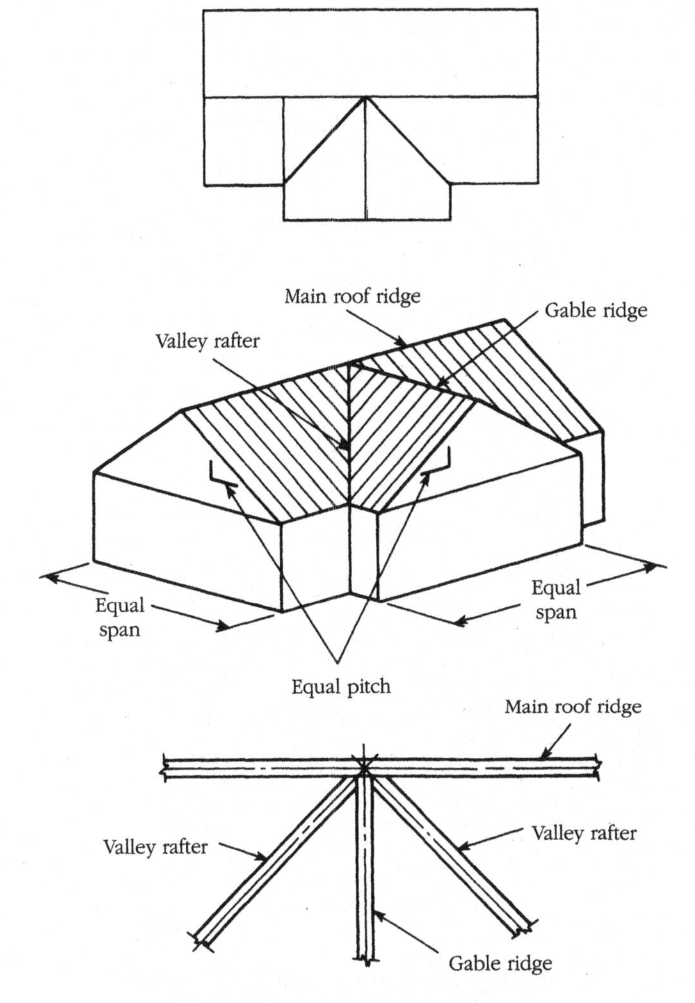

8-10 *View of a gable roof addition with equal pitch and equal spans.*

rafter of the main roof. The unit of run of a hip rafter or a valley rafter will therefore be identical.

Intersecting roofs of unequal spans

If the span of the roof of a dormer, addition, or wing adjoining the main structure is not the same as the span of the main building, the ridge lines will be at different heights. The treatment for this situation is seen in Fig. 8-11. One valley rafter is framed from the rafter plate of the secondary structure's roof to the ridge of the main structure's roof. The opposite valley rafter

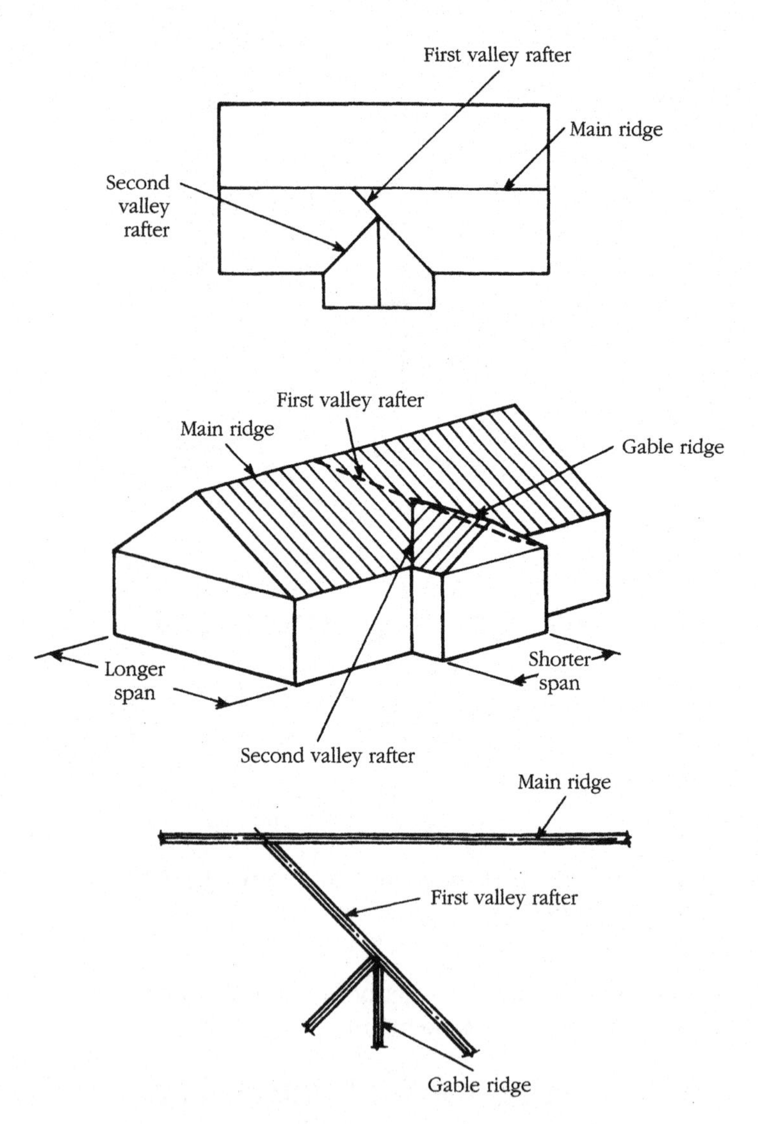

8-11 *Framing a secondary roof with a ridge line higher or lower than the ridge line of the main roof.*

will then be framed from the rafter plate of the secondary structure's roof and framed to the first valley rafter.

This configuration forms two right triangles, the first with a hypotenuse formed by the longer valley rafter (which is framed to the ridge of the main roof). Each of the two shorter sides of this right triangle are equal in length to the total run of the main roof's common rafter. The other right triangle formed has a hypotenuse consisting of the shorter valley rafter, with each of the two shorter sides equal to the total run of a common rafter of the secondary roof.

To determine the length of the valley rafters (line length), first remember that valley rafters and hip rafters are calculated in the same manner: thus, the unit run of a valley rafter equals that of a hip rafter, at 16.97. Next, consider the span of the secondary roof. If the secondary roof has a span of 20 feet, the total run of the shorter valley rafter would also be calculated in the same manner as a hip rafter would be—by adding half the span squared to half the span squared, then deriving the square root of the sum. With the numbers plugged in, the equation appears as:

$$\text{(span) } 20 \text{ feet} \div 2 = 10 \text{ feet}$$
$$10^2 + 10^2 = 200, \text{ so the square root of } 200 = 14.14$$

This means that the total run of the shorter valley rafter is 14.14 feet.

With a common rafter of the secondary roof having a unit of rise of 8 inches, the next step is to locate the 8-inch mark on the rafter table of the steel square and follow down to the second row of numbers, which provides hip and valley rafter bridge measures. The rafter table presents 18.76 as the corresponding bridge measure for a common rafter unit rise of 8.

Mathematically, by plugging these numbers into a proportional statement (previously discussed in this chapter), the line length can be expressed as the value of *X* in that statement:

$$16.97 : 18.76 :: 14.14 : X$$
$$16.97X = 14.14 \times 18.76$$

which further reduces to

$$16.97X = 265.27$$

Dividing 265.27 by 16.97 results in $X = 15.63$, so the length of the rafter would equal 15.63 feet.

For practical purposes, however, a shorter method of calculation does exist. Once the bridge measure has been derived as shown (18.76, in this case), multiplying it by half the span of the roof for which the valley is intended will provide the length of the valley rafter in inches; divide by 12 to arrive at the length expressed in feet.

Jack rafter layouts

Jack rafters are those portions of common rafters that are shortened to facilitate framing to a hip and or a valley rafter. For a framing of equal pitch, the jack rafter will always have the same unit rise as that of the common rafter. Another characteristic of jack rafters is that they are always framed on the same spacing on center as are the common rafters.

With this information, refer to Fig. 8-12: the shortest jack rafter shown becomes the hypotenuse of a right triangle, with each shorter leg measuring 16 inches. As before, the total run of this jack rafter is computed as the square root of 16^2 plus 16^2. (If the spacing on center were 24 inches, the total run of the jack rafter would be stated as the square root of 24^2 plus 24^2.) Knowing these characteristics of the jack rafter allows the application of previously discussed methods for determining various dimensions.

For example, if the unit of rise of a common rafter is 8, a jack rafter in the same roof would have a unit of rise of 8 as well; thus, the unit of length of a jack rafter would be expressed as the square root of 12^2 plus 8^2. Reduced, the statement means that for every 12 units of run, the jack rafter would be 14.42 units long. The numbers plugged into a proportional statement reads as follows:

$$12 : 14.42 :: 16 : X$$

Carrying out the calculations as before, the final answer reveals that the length of the shortest hip jack measures 19.23 inches. In other words, whenever the unit rise is 8 and the jacks and common rafters are 16 inches on center, the shortest hip jack will always be 19.23 inches long.

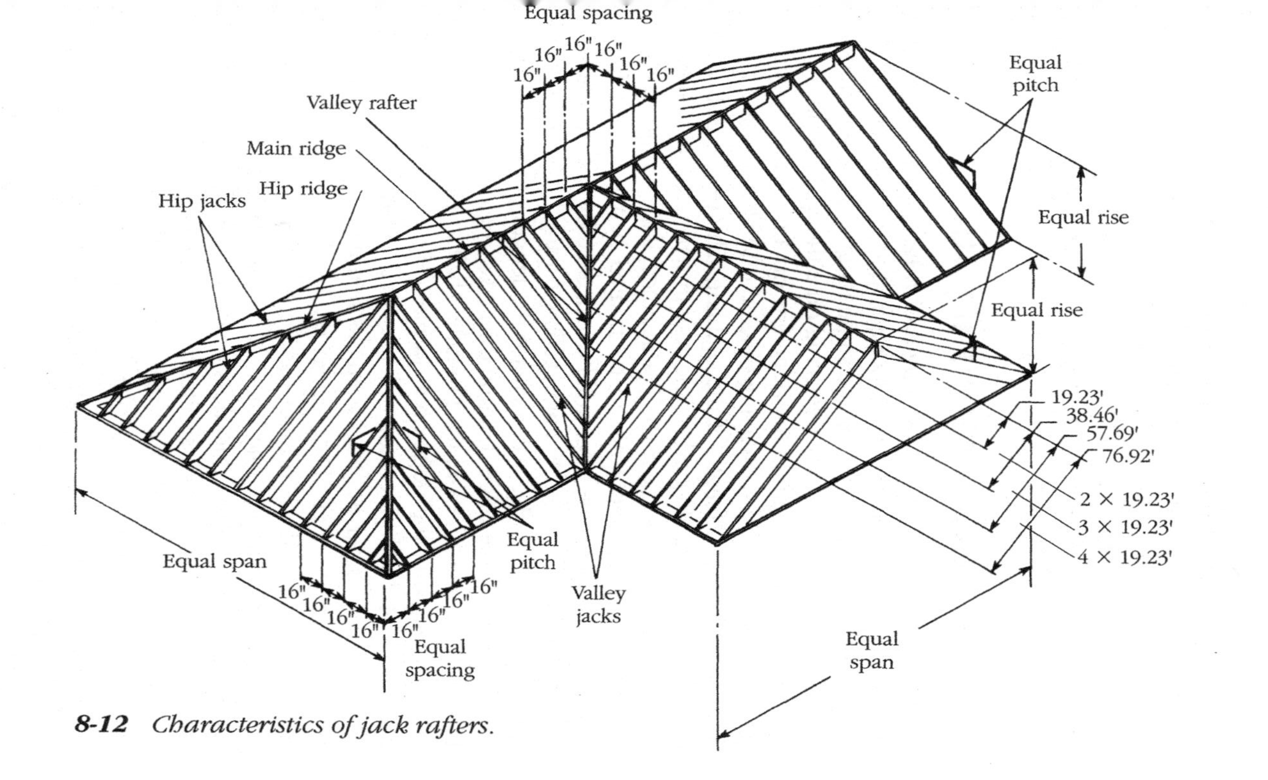

8-12 *Characteristics of jack rafters.*

Because the rafters are evenly spaced at 16 inches on center, 19.23 inches becomes a factor for determining the lengths of the successive jack rafters. With the shortest hip jack being 19.23 inches long, the next hip jack will be 38.46 inches long (2 × 19.23), the third hip jack will be 3 times 19.23 inches long and each additional hip jack's length will be determined by multiplying 19.23 by a multiplier one number greater than the multiplier used before it.

Ridge rafter layout

As opposed to calculating the length of ridge rafters on gable roofs, hip roofs are a bit more complicated. The length of a ridge rafter for a gable roof is the same length as the length of the building itself, plus any overhang on either gable end.

In theory, the length of the ridge line on a hip roof of equal pitch is calculated by subtracting twice the total run of a common rafter from the length of the building. The results of these calculations will be affected by the way in which the hip rafters are actually framed to the ridge. For instance, if the framing technique calls for the hip rafter to be framed between common rafters, the length of the ridge will be increased by half the thickness of a common rafter. Framing calling for the hip rafter to be framed against the ridge will increase the length of the ridge by the thickness of the ridge (half the thickness at either end), plus the 45-degree thickness of the hip rafter (half at either end).

The length of the ridge on a dormer (Fig. 8-13) is equal to the length of the dormer rafter plate added to the run (half the span) of the dormer. Subtract from this figure half the thickness of the inside member of the upper double header as a shortening allowance. If the dormer has no sidewalls, the length of the ridge is the same as the run (half the span) of the dormer. Again, subtract half the thickness of the inside member of the upper double header as a shortening allowance.

Calculating an addition's ridge length, with a span equal to that of the main building, involves two steps:

1. Adding the run of the addition (half the span) to the length of the rafter plate
2. Allowing half the thickness of the roof ridge on the main structure as a shortening allowance.

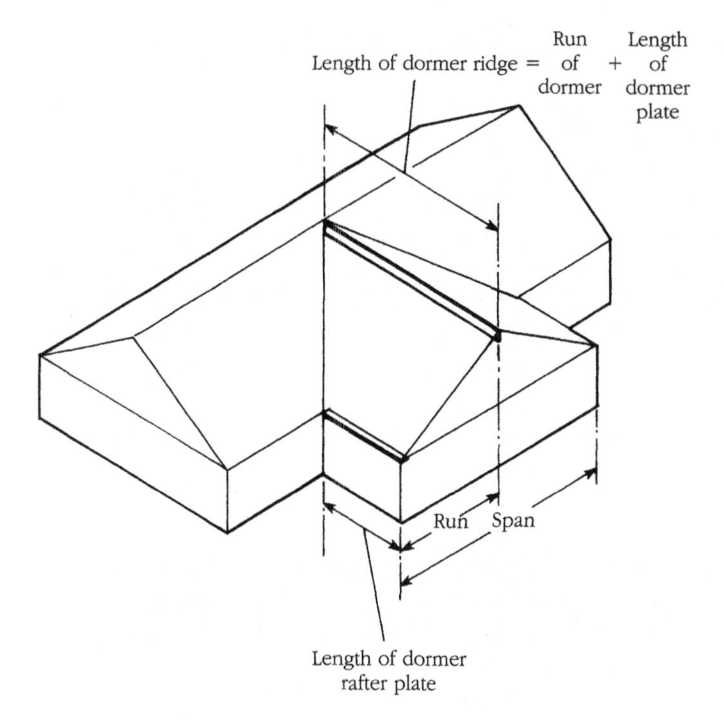

8-13 *A dormer ridge line.*

If the addition or wing has a span not equal to that of the main structure, the calculation of the ridge length will be affected by the construction method used to frame the ridge.

If, as earlier described, the ridge is framed with one long and one short valley rafter, the length of the ridge rafter will be equal to the length of the rafter plate of the addition, plus the run of the main structure. The shortening allowance in this case would equal half of the 45-degree thickness of the long valley rafter.

Another method entails framing the ridge of the addition to a double header set between double main roof common rafters. In this case, the length of the ridge would be equal to the length of the addition's sidewall rafter plate, plus the run of the addition. A shortening allowance of half the thickness of the inside member of the double header is then subtracted from the figure.

Framing shed roofs

As discussed earlier in this chapter, a shed roof can be thought of in terms of being the equivalent of half of a gable roof. Using this comparison, the total run of a shed roof should be equal to the span of the structure as opposed to half the span (as is the case with gable roofs). For the most part this is true, but, to be completely accurate, the width of the rafter plate on the higher of the two rafter end walls must be subtracted from the span for it to equal the total run. The width of the rafter plate is then included in the length of the overhang extending from the higher of the two end walls.

If a shed roof is to be used on a dormer projecting from the main structure's roof, the total run of a dormer rafter must be determined:

1. Calculate the difference between the unit rise of the dormer roof and the unit rise of the main roof.
2. Divide the height of the dormer sidewall or endwall (in inches) by the results of Step #1.

For example, assume a dormer with a sidewall height of 7 feet 6 inches (90 inches) and a unit rise of 2½ feet on the dormer roof. If the main structure's roof has a unit rise of 8, the math looks like the following:

Difference in unit rise between main roof and dormer roof

$$8 - 2\tfrac{1}{2} = 5\tfrac{1}{2}$$

Height in inches of dormer sidewall divided by 5½

$$\frac{90}{5\tfrac{1}{2}} = 16.36$$

With these numbers, the total run of the dormer rafter would be 16.36 feet.

To obtain the necessary stud lengths for a shed dormer's sidewalls, a proportional equation is again used. Using the previous example, the shed dormer has a dormer rafter with 2½ units of rise for every 12 units of run. The roof of the main structure has common rafters with a rise of 8 units for every 12 units of run. If the plans called for the studs to be spaced 12 inches on center, a common difference could be computed as it was with jack rafters. By subtracting 2½ inches from 8 inches, 5½ inches would be the length of the shortest stud and is also the multiplying factor used to determine the lengths of the remaining studs. The next shortest rafter would be twice 5½, or 11 inches in length. The third rafter would be three times 5½, or 16.5 inches long. In order to modify this common difference for a situation in which the stud spacing is 16 inches on center instead of 12 inches on center, a proportional equation with X representing the length of the shortest stud would be employed as follows:

12 (the stud spacing in the example) is to 5½ (the common difference in the example) as 16 (the actual stud spacing) is to X (the value of the common difference corresponding to the actual stud spacing)

$$12 : 5\tfrac{1}{2} :: 16 : X$$

$$12X = 5\tfrac{1}{2} \times 16$$

$$12X = 88$$

$$\frac{88}{12} = 7\tfrac{1}{3}$$

The length of the shortest stud, which is also the common difference, would be 7⅓ inches. The next longest stud would be 14⅔ inches (7⅓ × 2), the third stud would measure 22 inches and so on.

Roof sheathing

As with subflooring or wall sheathing, roof sheathing forms part of the framing of the structure and is laid across and attached to the rafters. Whether the sheathing is to be material sheets or boards, determining the quantity of sheathing required is accomplished by calculating the total square footage to be covered.

Formulas for calculating roof area

Figuring the area of a shed roof is the easiest, since it only requires multiplying line length from the eave edge to the ridge by the span.

A gable roof presents the next most simple example of calculating roof area. Multiplying the distance from the eave edge to the ridge by twice the length of the ridge line yields the total area of the roof. If, for example, a gable roof is 15 feet from eave edge to ridge with a ridge line of 40 feet, then $15 \times (2 \times 40) = 1200$ square feet.

If sheathing boards are to be used, the quantity of material (in board feet) is the same as the total area. Since boards used for sheathing are normally of 1-inch stock, once the previous calculations have been done, 1200 board feet of sheathing would be needed. Adding a 15-percent waste factor would bring the total up to 1380 ($1200 \times 0.15 = 180$) board feet.

A 4 by 8 sheet of sheathing material equals 32 square feet of surface, and dividing the total square footage of the roof by the square footage of a single sheet of material will yield the number of sheets of material needed to sheath the roof (in this case, $1200 \div 32 = 37\frac{1}{2}$). With roof sheathing, a waste factor of 15 percent should be added ($37\frac{1}{2} \times 0.15 = 5.62$); consequently, a rounded total of 43 sheets would be appropriate.

If a hip roof is on a square building, the geometric formula for determining the area of a triangle becomes very useful. If one separates the roof into four equal triangles, the area of one triangle can be calculated with the following formula:

$A = \frac{1}{2} bh$ (read as "area = half the triangle's base times its height")

In Fig. 8-14, the triangle's base is 20 feet, its height 15 feet. Utilizing the geometric formula from earlier reveals the following:

$$A = \tfrac{1}{2} bh$$

$$A = (20 \times 15) \div 2$$

$$A = 300 \div 2$$

$$A = 150$$

So the area of one triangle is 150 square feet. Since the roof comprises four triangles of equal dimensions, multiplying the area of one triangle by 4 will equal the total area of the roof.

$$150 \times 4 = 600 \text{ square feet}$$

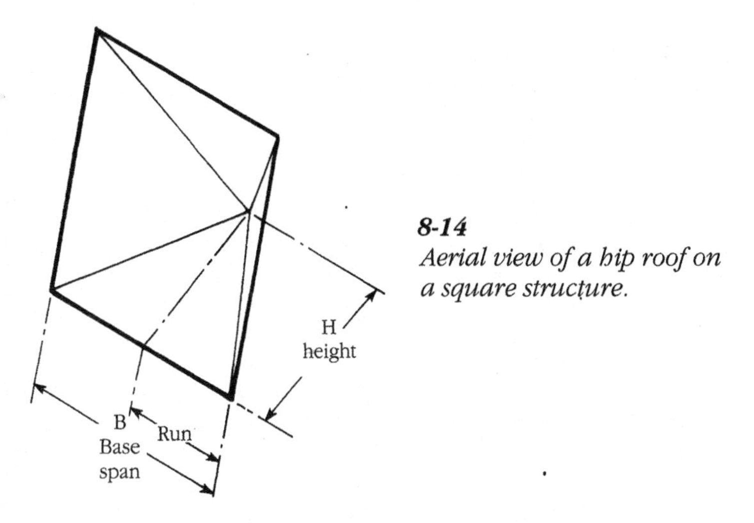

8-14
Aerial view of a hip roof on a square structure.

To calculate the area of a hip roof on a rectangular structure (Fig. 8-15) proceed as follows:

1. Multiply the length of the eave edge along the short side of the rectangle (A) by the height of the triangle (B) which is formed by the roof on that side (measuring from the eave edge to the ridge).
2. Determine the length of the eave edge along the long side of the rectangle (C). Add this figure to the length of the ridge line (E) and multiply the sum by the height of the line measure from eave edge to ridge line (D).
3. Adding the results of Step #1 and Step #2 will equal the total area of the roof (Area equals A times B plus the product of D times the sum of C plus E).

$$Area = (A \times B) + D(C + E)$$

Mansard roofs (Fig. 8-16) are calculated with the following formula:

$$A = C(D + A) + E(F + B) + (A)(B)$$

The area of a gambrel roof (Fig. 8-17) is derived by multiplying the sum of the line lengths of the two roofing segments which descend from the ridge line by twice the length of the ridge. Mathematically, this formula is

$$area = 2(C) \times (A + B)$$

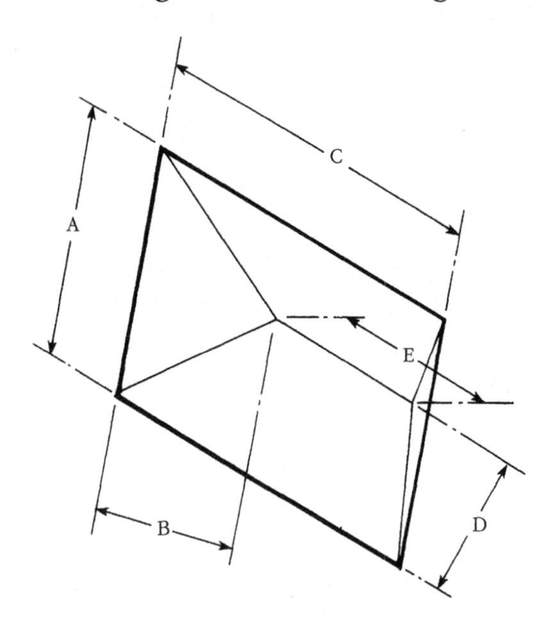

8-15 *Measuring a hip roof on a rectangular structure.*

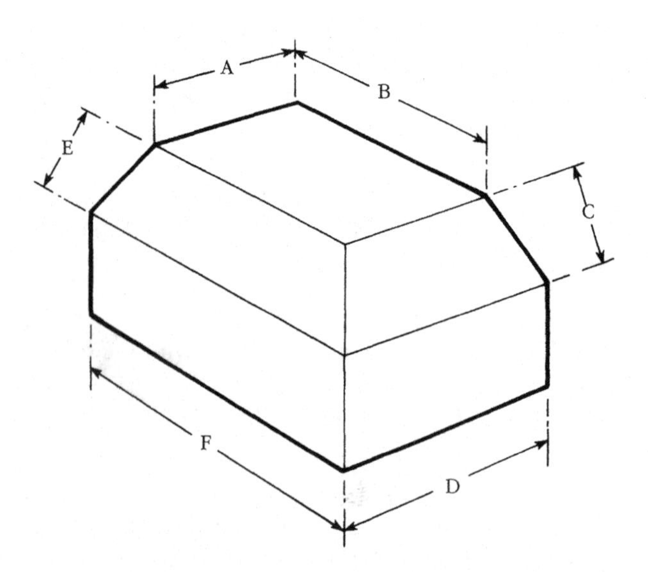

8-16 *Example of a mansard roof.*

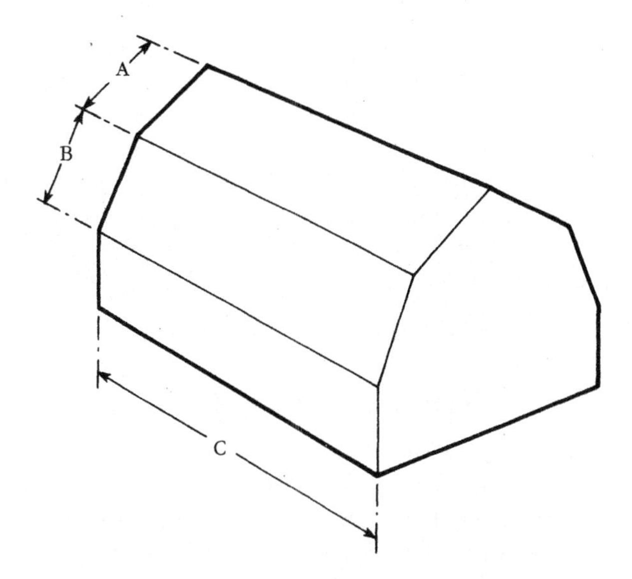

8-17 Calculating the area of a gambrel roof.

Slope factors

An alternate method for determining a roof's area is to use the slope (in inches) of a roof to convert the ground area occupied by the structure into roofing area. Table 8-1 provides numbers

Table 8-1
Slope multipliers used for converting ground area to roof area.

Roof slope	Multiplier
2/12	1.11
3/12	1.13
4/12	1.15
5/12	1.18
6/12	1.22
8/12	1.30
10/12	1.40
12/12	1.51
16/12	1.72
18/12	1.90

corresponding to common roof slopes. These numbers are used as multipliers by which the ground area of a structure is multiplied in order to obtain the area of the roof.

For example, if a structure measures 50 feet by 30 feet and has a gable roof with a slope of 6/12, consult the table for the multiplier corresponding to a slope of 6/12. Once the appropriate multiplier has been located (1.22), the math required to calculate the area of the roof is as follows:

$$30 \text{ feet} \times 50 \text{ feet} \times 1.22 = 1830 \text{ square feet}$$

Roof coverings

Applied on top of the sheathing, the exterior roof finish material serves two purposes: aesthetics and protection. The type of finish roofing material chosen must compliment the architecture of the structure while providing the building with protection against the elements—including but not limited to wind, sun, rain, snow, ice and fire. Roof coverings are as follows.

Asphalt shingles

Asphalt shingles are the most commonly used material for residential roofing in the United States today and are made in a variety of styles. The most popular is the square butt strip shingle, elongated in shape and available with three tabs, two tabs, or one tab (without cutouts).

Typical asphalt square butt strip shingles are made of a base mat of organic material (cellulose fibers) or inorganic material (glass fibers) saturated and coated with asphalt and surfaced with ceramic-coated opaque mineral granules. The asphalt provides waterproofing qualities, while the mineral granules protect the shingles from the sun's drying rays and give the shingles color and added fire protection.

Most shingles are available with strips or spots of a factory-applied self-sealing adhesive that is a thermoplastic material activated by the heat of the sun after the shingle is on the roof. Exposure to the sun's heat bonds each shingle securely to the one below it for greater wind resistance. During the spring, summer, and fall, this self-sealing action usually takes place within a few days of the installation. In winter, the self-sealing action varies depending upon the geographic location, roof slope, and orientation of the house on the site.

Interlocking shingles are also available and designed to provide immediate resistance to strong winds. These shingles come in various shapes and with various types of locking designs that provide a mechanical interlock on the roof.

Today's generation of dimensional or architectural asphalt shingles has elevated the art of roof design, especially in the residential re-roofing and new construction markets. Thicker, stronger, and more natural-looking than typical three-tab shingles, dimensional shingles add lines of depth and distinction to homes, increasing curb appeal. What is more, with the addition of shadow lines and random, laminated tabs, architectural shingles can capture with uncanny accuracy the old-fashioned warmth and elegance of roofs constructed of wood shakes or shingles and slates. Many of these shingles carry warranties ranging from 20 to 40 years. They are fungus-resistant and have a fire rating from Class C to Class A.

Wood shingles

Wood shingle and shake roofs, from an historical perspective, could be considered the most American of all roofing materials. The seemingly endless supply of forested land on the newly settled continent made wood roofs very prevalent in Colonial times—an era in which Europe's most common roof coverings were slate, tile, and thatch. Wood species as diverse as oak, eastern white cedar, pine, hemlock, spruce and cypress were all used, depending on local availability.

During the late 1800s, the more simple utilitarian function of wood shingles gave way to a full spectrum of architectural and artistic expressions. Styles known variously as Carpenter Gothic, Queen Anne, and Shingle-style appeared on the scene. This new breed of domestic architecture was often defined by elaborate exteriors, clad entirely with wood shingles, which most often were western red cedar.

Today, most cedar shakes are machine split on one side and resawn on the other, with the split side laid exposed to the weather. According to the Cedar Shake and Shingle Bureau, (the recognized authority in the grading and specifications for cedar roofing) cedar shakes comprise the biggest share of the total mill production of cedar roofing material.

Under certain municipal and regional building codes, wood shakes, and shingles are not permitted unless treated with a fire-retardent agent. A lifetime warranty on fire-retardent cedar shakes and shingles is now available with a warranty that ensures the fire protection will last for the life of the roof. Such certifications are only awarded to fire-treated shakes and shingles that have successfully passed rugged 10-year weathering tests. This allows them to meet all national building code specifications for roofs requiring Class B or C fire ratings.

Slate

Slate differs principally from other stone because of its natural cleavage construction. These cleavages allow it to be easily split in one direction to desired thicknesses and strengths.

A slate roof is truly a custom job and gives a building a unique character all its own. Successive beds in a quarry may consist of various colors of slate, ranging from gray to green to purple to black and a spectrum of colors in between. Generally, Pennsylvania slate colors are blue-grey, blue-black, and black. Buckingham and other Virginia slates are generally blue-grey to dark grey, with micaceous spots on the surface that produce an unusual luster. Vermont slates can be light grey, grey-black, unfading and weathering green, unfading purple, and variegated mottled purple and green. Unfading red slates are found only in Washington County, New York, and are the most costly. Also, a lustrous unfading black slate of exceptional strength was long quarried in the town of Monson, Maine, but is no longer sold as roofing slate.

As a roofing material, slate is long-lasting and extremely durable. One of the most permanent roofing materials available, certain types of slate have a longevity of over 175 years. No man-made roofing material can make that claim. As it comes in a wide range of colors and textures, slate can add aesthetically to the overall beauty of a home (a slate roof adds appreciably to the value of a house, too). Because it is made of durable stone, slate needs little ongoing maintenance—no painting, coating, or cleaning required. And slate is practically immune to changes in the weather.

Clay tile

First appearing in the Bronze Age, sun-baked tiles have been found in the archaeological digs of ancient Crete. Today, clay tile is still among the most popular roofing materials in Europe, where homes and centers of commerce are designed to last for generations. The simplicity of form and shape makes traditional tapered Mission tiles ideal conveyors to funnel and shed water from pitched roofs. Modern extrusion and pressed-formed processes and high-tech gas-fired kilns have replaced the primitive method of shaping clay tiles over human thighs and then either baking them in the sun or wood-fired "beehive" kilns. And with these advances in manufacturing have come tremendous improvements in performance, quality, and product diversity.

Many architectural details are possible only with the physical attributes of tapered Mission clay tiles. Deep shadow contrasts are created with high-profile clay tiles. Historically accurate accents are available through boosting and mudding application techniques. Traditional hip stacks and various methods of eave closure, including clay "birdstops" and mortar fill, are eye-appealing enhancements. Scratched or brush-finished textures applied to tops and pans, and even one-piece "S" shape tiles, can simulate the appearance of age from the first day of installation. Architects consistently specify tapered "mission" in order to turn tightly radiused roofs and turrets, a feat not possible with many contemporary roof tiles.

Fiber cement shingles

Fiber cement roofing is a highly desirable alternative because of its close resemblance to natural products. Varying from manufacturer to manufacturer, some resemble clay tiles, others imitate thick cedar shakes by being randomly formed, and still others take on the look of slate in form, color, and appearance.

The base for these products is cement, combined with a variety of other natural and synthetic fibers (such as wood fibers, perlite, cellulose, recycled fly-ash, or fiberglass materials). Iron oxides, resins, and other materials are added for color definition, finish, and strengthening. These products also meet lightweight roof covering requirements, which eliminates the need for additional rebracing of the existing framing structure. Many fiber cement products can be installed directly over spaced sheathing, which is the most likely roof substructure if the old material was wood or shingles.

Fiber cement roofing is warranted for long-life expectancy (30 to 50 years) and most manufacturers of these materials offer a transferable warranty, which enhances the homes's value when re-selling. All these products are fire-safe, meeting the standards set by the Uniform Building Code for Class "A" or "B" fire ratings.

In addition to the re-roofing market, these products are gaining popularity in the new construction market as well. This is largely attributed to the desire of many developers—because of increased building costs—to upgrade materials used for new housing.

Thermoplastics

Thermoplastics are excellent roofing materials that can be molded into authentic-looking wood shake and slate facsimiles. Actually, thermoplastics can be molded into so many different shapes and profiles that people have trouble telling the plastics from natural products. Looking at the benefits of these plastics over other roofing materials, fire protection is excellent, since with an ignition temperature in excess of 1100 degrees F., these roofing resins do not often burn. Thermoplastic resin roofing systems are easy to work with. Each manufacturer has their own interlocking system, but the panels themselves are light (65 to 90 pounds per square). The panels can be cut with a circular saw, and the systems are completely walkable. Panels are different sizes and range from 20 pieces to 12 pieces per square. Some other characteristics of these shingles include wind and water resistance, energy efficiency, recyclability, and impact resistance.

Metal roofing

Today, the reasons for choosing metal roofing are both obvious and surprising. Metal roofing allows the owner or designer to make a design statement. When the roof is high-pitched and part of the integral building design, the architectural possibilities of metal roofing are not attainable with any other material. When the roof is low-pitched and serves solely as a functional water barrier, the weatherproofing capabilities of metal roofing exceed any other materials available.

The two basic types of metal roofing are structural and architectural. Structural metal roofing has load carrying capabilities and does not require a substructure for support. Structural metal roofing, depending upon profile and material thickness, can be applied on very low slopes. The industry minimum standard pitch for structural steel roof systems is ¼ inch in 12 inches. Other materials, aluminum and copper, as well as various alloys and plated composites, have structural limitations to be considered. Structural standing seam roofing is comprised of interlocking panels, varying from 12 to 24 inches in width that are designed to form a continuous waterproof membrane across the roof. The configuration of the panels is either a trapezoid-shaped rib, 2 to 3 inches high, or a square, vertical, narrow rib 2 to 3 inches high. Both of these basic panel configurations rely upon the shape of the panel to provide the structural capability to span between structural members.

The standard for structural panel free span is 5 feet. There are other factors to be considered, such as design load requirements and existing structural member locations (in re-roof. applications), that may alter the free span capability, but 5 feet is the most common. Having a free span capability means that no substructure is required to support the metal roofing panels.

Architectural panels are non-structural and must be supported by a substructure. In most cases, the substructural support is plywood decking. Metal decking or fire-retardent plywood is used when non-combustibility is required. Architectural panels, due to their design, are water shedders as compared to the structural standing seam panels, which are water barriers. The minimum roof slope for the architectural panels is generally considered to be 3 inches in 12 inches. Roofing felts are usually installed between the panel and the substructure, acting as an additional moisture protection.

The use of metal roofing forms an accent to masonry, wood, stucco, glass curtain walls, marble, and granite. The various profiles available in metal panels make project adaptation from large to small very easy. The 18- to 24-inch-wide panel configurations are best suited for large commercial structures, while 8- to 12-inch-wide profiles are most often seen onsmaller structures such as residential or smaller commercial buildings.

Determining quantities of roofing material

Roofing material is generally supplied in standard units of measure referred to as squares. That is, a square of roofing material will include enough material to cover an area 100 square feet in size. Just as with sheathing, deriving the number of squares of material needed to cover a given roof requires that the total area of the roof be known. Calculating the total area of different styles of roofs is discussed in detail in the section of this chapter dealing with sheathing.

Once the total square footage of roof area to be covered is known, divide that number by 100 (1 square = 100 square feet) to determine the number of squares necessary. For example, if the total square footage of a roof equals 2700 square feet, then 27 squares of material will be needed for that roof.

$$\frac{2700}{100} = 27$$

To this figure, add a waste factor of 10 percent, rounded to the next highest square, or in this case 3 squares.

$$27 \times 0.10 = 2.7 \text{ (which equals 3 when rounded up)}$$

For roofing projects that include valleys, one square of roofing material should be added to the total for every 100 linear feet of valley area on the roof. If there are 180 linear feet of valley on the roof, the estimate would be rounded up to two additional squares of material.

Actual coverage area per square will vary depending on the amount of weather exposure per course of material and the material itself. Table 8-2 provides information based on typical four-bundle squares of various size asphalt shingles at differing weather exposures.

Notice that, with a 4-inch exposure, a square of 16-inch No. 1 blue label or No. 2 red label shingles will actually cover 80 square feet. Using this information, the 2700 square foot roof mentioned earlier would require 34 squares of material.

$$\frac{2700}{80} = 33.75 \text{ (rounds up to 34)}$$

Table 8-2 Coverage area of shingles at various weather exposures.

Length and thickness	Approximate coverage of one square (4 bundles) of shingles based on weather exposures								
	3½"	4"	5"	5½"	6"	6½"	7"		7½"
16" × 5.2"	70	80	90	100	-	-	-	-	-
18" × 5½–¼"	-	72½	81½	90½	100*	-	-	-	-
24" × 4/2	-	-	-	-	73½	80	86½	93	100*

NOTE: Maximum exposure recommended for roofs.

The same principle used as the basis for Table 8-2 applies to shake roofing as well. Table 8-3 supplies the amount of coverage per square of shakes which can be expected at differing weather exposures.

Roll roofing

Beneath asphalt shingles as well as some other types of roofing, 15 pound felt underlayment is commonly used as additional weather protection and preparation for the final roofing material.

A standard roll of 15-pound roofing felt measures 3 feet by 12 feet (432 square feet). However, due to the manner in which this product is applied, with lapped joints, a roll of felt will actually cover 400 square feet. Consequently, the number of rolls of felt for a 2700-square-foot roof would be calculated as:

$$\frac{2700}{400} = 6.75 \text{ (rounds up to 7)}$$

It is important that the contractor be comfortable with the various methods and calculations discussed in this chapter. However, with the advances in calculators and particularly in computers, performing these calculations by hand is becoming less and less common. A major productivity feature of advanced software systems is the "work package"—allowing the roofing contractor to take off a roof's numerous items just by entering its dimensions. To take off a 4-ply built-up roof, for example, enter the roof's length, width, and slope. The work package will automatically estimate labor, material, and equip-

Table 8-3 Shake coverage area at various weather exposures.

Shake type, length and thickness	Approximate coverage (in sq. ft.) of one square when shakes are applied with an average ½ inch spacing at the following weather in inches (d):				
	5	5/12	7½	8½	10
18" × ½" handsplit and resawn mediums (a)	-	55(b)	75(c)	-	-
18" × ¾" handsplit and resawn heavies (a)	-	55(b)	75(c)	-	-
18" × ⅝" tapersawn	-	55(b)	75(c)	-	-
24" × ⅜" handsplit	50(e)	-	75(b)	-	-
24" × ½" handsplit and resawn mediums	-	-	75(b)	85	100(c)
24" × ¾" handsplit and resawn heavies	-	-	75(b)	85	100(c)
24" × ⅝" tapersawn	-	-	75(b)	85	100(c)
24" × ½ tapersawn	-	-	75(b)	85	100(c)
18" × ⅜" straight split	-	65(b)	90(c)	-	-
24" × ⅜" straight split	-	-	75(b)	85	100(c)

Use as supplemental with shakes applied not to exceed 10″ of 15″ starter-finish course weather exposure.

(a) 5 bundles will cover 100 sq. ft. roof area when used as starter-finish course at 10″ weather exposure; 7 bundles will cover 100 sq. ft. roof area at 7½″ weather exposure; see (d).
(b) Maximum recommended weather exposure for 3-ply roof construction.
(c) Maximum recommended weather exposure for 2-ply roof construction.
(d) All coverage based on an average ½″ spacing between shakes.
(e) Maximum recommended weather exposure.

ment costs for all the felts, asphalt, surfaces, fasteners and any other items related to that roof. Based on the length of gravel stop, the intelligent work package will count the number of joints required, add the proper amount of material for the lap, and round up to the next 10-foot piece.

From this take-off, the work package will calculate estimated labor and material costs for numerous items, including insulation, fasteners, felts, and surfaces, as well as taking into account waste factors and transferring information into a complete bill of materials and a detailed field productivity report.

Review questions

1. Using the diagram in Fig. 8-18, what would be the length of the bottom of the rafter from pt L to pt M? Pitch of LM = 5 in/12 in
2. An estimator is figuring out how many bundles of shingles to order for the roof on a garage. The garage is 24′ × 24′, with a single roof ridge down the middle, plus a 2-foot overhang all around. The roof slope is 6″ by 12.″ Three bundles of shingles are required for each 100 square feet of roof. How many bundles of shingles should be ordered?
3. A customer would like a bonus room to be added to an existing home. The new room's dimensions are 26′ × 22′ with an 8′ ceiling. The ridge of the roof is centered over the 22′ wall. The height of the roof is 5 feet and has a 2-foot overhang. How many sheets

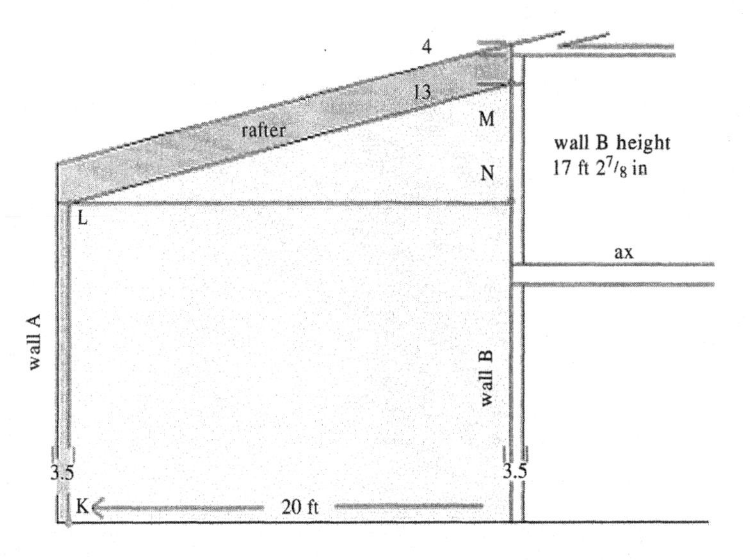

of plywood are needed to cover the roof with a 2′ overhang?

4. You need to roof a house with composite shingles. Each side of the roof is 20 feet by 40 feet. Each package of shingles covers 12 square feet (ft^2). How many packages of shingles are required?
    a. Each package of shingles costs $7 plus another $12 to have it put on the roof (for labor and other materials such as nails). How much will the entire job cost?

9

Math used for finishing interiors

Finishing the interior of a structure as discussed in this chapter will include such areas as millwork, stair construction, walls, ceilings, floor coverings, and painting.

Stair construction

All of the many configurations of stairs share two common components: the treads and the stringers supporting them. Stairs that consist only of treads and stringers may usually be found in casual environments, while in more formal settings risers connecting the inside edge of one tread to the outside edge of the tread above are often added. In this chapter, most of the discussion concerning stairs deals with these three elements.

Laying out stairs

When laying out a stairway, the unit run and unit rise (Fig. 9-1) of the stairs must be derived from the total rise of the stairway. Total rise is the total vertical distance that the stairs must bridge from the lowest point of the stairs to their highest point in order to connect the two floors of different elevation.

Stairways are normally laid out after the subflooring has been laid but before the finish flooring is installed. Once the vertical distance the stairs are to span has been measured, the thickness of any finish flooring to be installed on either the upper or lower level must be added to the vertical distance to accurately obtain the total rise.

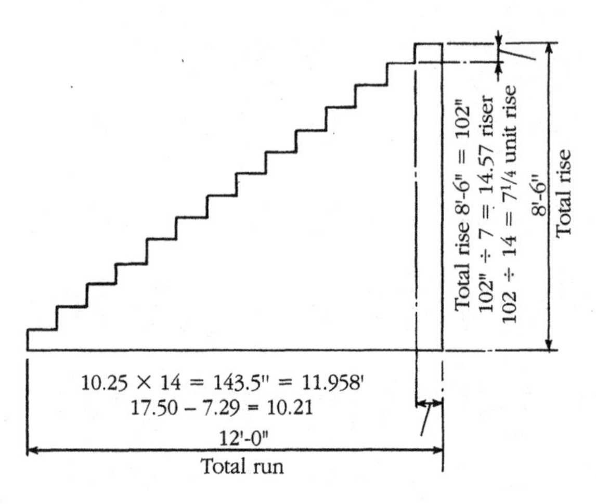

9-1 *Components of a stairway.*

Architects and contractors employ two rules of thumb when laying out stairs:

- The average unit rise should be approximately 7 inches.
- The sum of the unit rise and unit run should equal approximately 17½ inches.

Keeping these two rules in mind, and assuming the total rise of a planned stairway equals 8 feet 6 inches, use the following procedure to determine unit run and unit rise:

1. Convert the total rise into inches. 8 feet 6 inches = 102 inches.
2. Calculate the number of risers needed for the stairway by dividing the total rise by 7 (rule of thumb #1): 102 / 7 = 14.57.
3. Disregard any fraction resulting from the calculation in Step #2 and round to a whole number; in this case, use 14.
4. To determine the unit rise, divide the total rise by the number of risers: 102 / 14 = 7.29 inches or roughly 7¼ inches.
5. By subtracting the unit rise (7.29) from the 17.50 inches stated in the second rule of thumb, you will get the unit run. 17.50 – 7.29 = 10.21 or roughly 10¼.

Thus, the unit rise is approximately 7¼ inches, and the unit run is approximately 10¼ inches—totaling the 17½ inches stated in the second rule of thumb.

Calculating the total run of a stairway

Calculating the total run of a stairway involves multiplying the number of whole treads times the unit run. However, the number of whole treads will vary with the technique employed to secure the upper end of the stairway to the well header.

If a complete tread is at the top of the stairs running level to the finish floor (Fig. 9-2), then the total number of whole treads is equal to the number of risers. Because there are 14 risers in this example, there would also be 14 treads. The total run of a stairway with these dimensions would equal:

$$14 \times 10.25 = 143.50 \text{ inches} = 11.96 \text{ feet (rounded to 12)}$$

If the finish flooring of the upper level takes the place of a tread at the upper end of the stairway (Fig. 9-3), the number of whole treads is reduced by 1 to a total of 13. The total run of the stairway is now 13 times the unit run:

$$13 \times 10.25 = 133.25 \text{ inches} = 11 \text{ feet } 1 \text{ inch}$$

For situations in which less than a whole tread is present at the upper end of the stairway, the total run would be calcu-

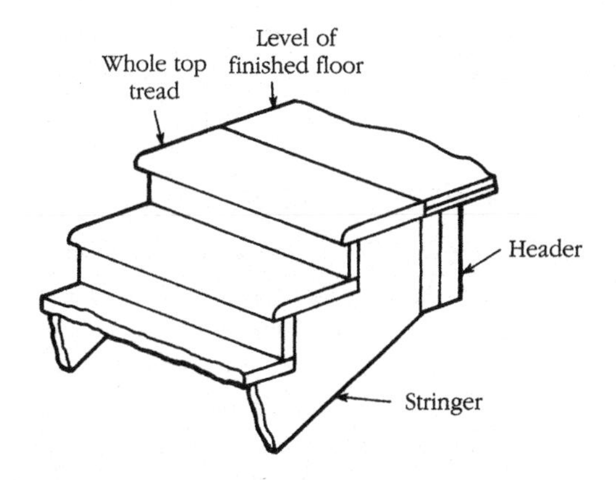

9-2 *Stairway construction anchoring a whole tread at the uppermost tread location directly to the well head.*

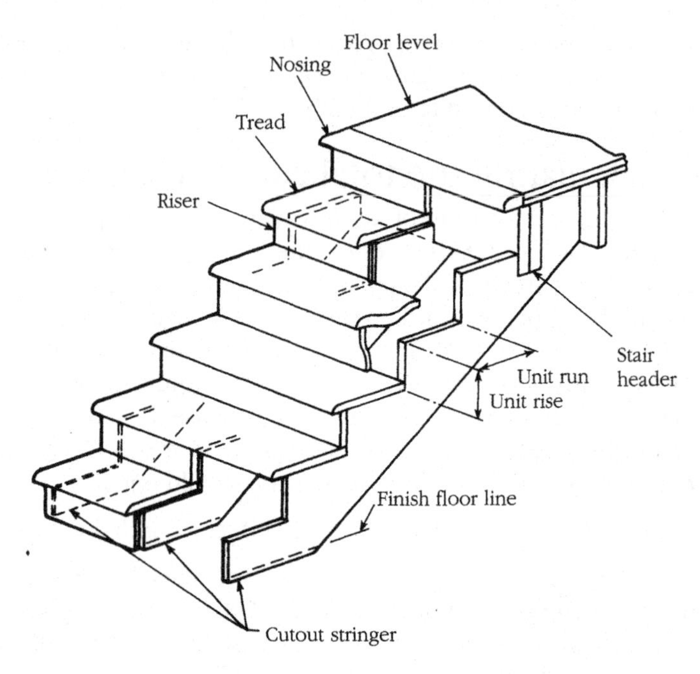

9-3 *Stairway construction where the finish flooring serves as the top tread.*

lated by multiplying the unit run by 13 and then adding the actual run of the partial top tread to the total. For example, if the top tread had an actual run of 5 inches, the calculations would be as follows:

$13 \times 10.25 = 133.25 + 5 = 138.25$ inches $= 11.52$ (or $11\frac{1}{2}$) feet

Fabricating stringers

Prior to cutting stringers for a stairway, one must calculate the length of a piece of stock from which the stringer will be cut. To obtain this number, the stairway must be viewed geometrically. The total rise and total run of the stairway are two lines, which seen geometrically form two sides of a triangle for which the stringer will form the hypotenuse. Because the lengths of the two sides are known (8 feet 6 inches and 12 feet), the hypotenuse can be calculated by determining the square root of $8.5^2 + 12^2$.

$$8.5^2 = 8.5 \times 8.5 \text{ or } 72.25$$

$$12^2 = 12 \times 12 \text{ or } 144$$

$$72.25 + 144 = 216.25$$

The square root of 216.25 = 14.71

Note: As shown in Appendix B, the decimal fraction 0.42 is roughly equivalent to 5/12, which in this instance translates to 5 inches.

So the length of the stringer stock would be approximately 14 feet 5 inches.

Locating the unit run along the tongue of a steel square and the unit rise on the body, plotting the position of each riser and tread becomes an easy chore—with one exception. Prior to defining the unit rise of the bottom riser, one must make a correction referred to as *dropping the stringer*, to account for the aggregate thickness of the treads. When the base of the stringer is to rest on the subflooring, the unit rise of the first riser (and only the first riser) is decreased by the thickness of a tread, minus the thickness of the finish flooring. If the base of the stringer is to rest on a finish floor instead of subflooring, the unit rise of the first riser would be decreased by the thickness of a tread. In the previous examples, where the unit rise equalled 7¼ inches, if the base of the stringer is to rest on a finish floor and each tread measures ⅛ of an inch in thickness, then the unit rise of the first riser would actually equal 7⅛ inches.

Wall construction materials

Applied to the wood framing on the inside of a structure are usually one of two categories of construction materials: lath-and-plaster or drywall. Although still used, lath-and-plaster has become less popular over the years, primarily due to the labor expense associated with it as well as the drying time required by the product and the specialized skill involved in applying the plaster. Dry-wall, on the other hand, has increased in popularity, not only because it is generally less expensive to install, but also due to the variety of dry-wall materials and the diversity of their usage.

Types of dry-wall material

Some of the more popular types of dry-wall include plywood, hardboard, gypsum board, and wood paneling. Each type of dry-wall possesses different combinations of characteristics both desirable and undesirable for a given installation. However, with the variety of dry-wall materials available, finding a type of dry-wall that will meet particular job requirements should not pose a problem.

Of the many different dry-wall materials manufactured today, gypsum board has become a very versatile product, offered in an array of compositions suited for various specific purposes.

Regular or standard core gypsum board Usually manufactured in 8-, 10-, 12-, and 14-foot panels, the board length will depend on the application and panel thickness. The standard width of a gypsum board is 4 feet, which is compatible with the framing of studs or joists spaced 16 inches and 24 inches on center (O.C.) and room heights of 8 feet. But, because of the popularity of 9-foot ceilings in some residential and commercial buildings, some manufacturers are making a 54-inch wide board. With these additional 6 inches of width, two horizontally placed pieces fit perfectly on 9 foot walls with no extra cutting, taping, or finishing. Other lengths and widths of gypsum boards are available from manufacturers on special order.

Regular gypsum boards are generally available in six thicknesses:

- *¼ inch.* A lightweight, low cost utility gypsum panel used as a base layer for improving sound control in multi-layer partitions and in covering old wall and ceiling surfaces. Also, it is used for forming curved surfaces with short radii.
- *5⁄16 inch.* A gypsum board usually employed in mobile homes to keep weight to a minimum.
- *⅜ inch.* A lightweight panel applied principally in repair and remodel work over existing surfaces. Also used frequently in the double-layer wall system.
- *½ inch.* Used in single-layer application in both new construction and remodeling. It can also be used in double-layer systems for greater sound and fire ratings.
- *⅝ inch.* Is recommended for the finest single-layer drywall construction. The greater thickness provides increased

resistance to fire exposure, transmission of sound, and provides higher rigidity.

- *¾ and 1 inch.* Known as coreboards, are used in solid drywall partitions, shaft walls, stairwells, chaseways, area separation walls and corridor ceilings.

Regular gypsum boards have a paper covering on each side and on the edges. The backs of the boards are surfaced with a gray liner paper, while the facing is generally a light gray manila paper extending over the long edges. The surface is smooth and will take a wide variety of finishes. The boards may be applied in one or more layers directly to wood framing members, to steel studs or channels, or to interior masonry surfaces.

Single-layer or single-ply gypsum board construction is the most commonly used system in residential and light commercial construction (Fig. 9-4). The single-ply uses just one layer of gypsum drywall and is usually adequate to meet many fire resistance and sound control requirements.

Multi-ply or -layer construction consists of a face layer of gypsum applied over a base layer of gypsum board directly attached to the framing members. This construction can offer greater strength and higher resistance to fire and to sound

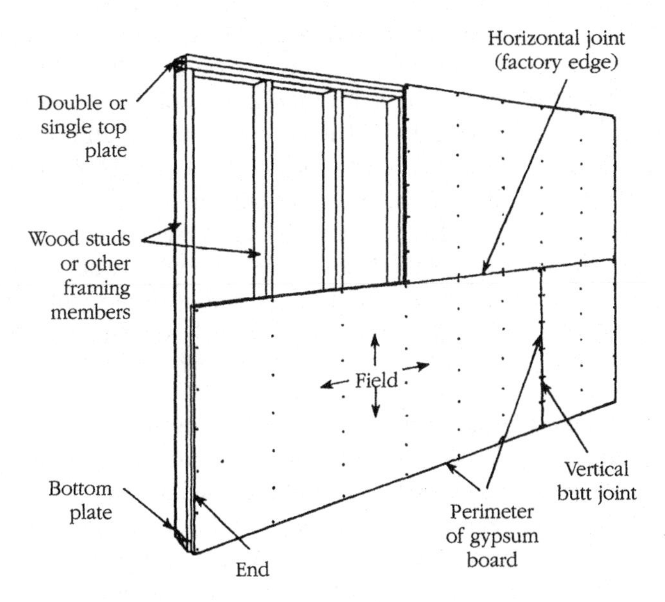

9-4 *A single-layer or single-ply gypsum wall.*

transmission than single-ply applications. Double-ply construction when adhesively laminated is especially resistant to cracking and provides a very fine and strong wall. These adhesively laminated constructions are highly resistant to sag and joint deformation.

Type-X gypsum board This board combines all the advantages of standard panels with additional resistance to fire exposure—the result of a specially formulated core containing special additives that enhance the integrity of the core under fire exposure. When used on both sides of a load-bearing wood stud partition, a ⅝-inch type-X board has a 1-hour fire rating, while a ½-inch type-X board provides a 45-minute assembly rating. Fire-resistant gypsum panels are generally available in 4-foot widths and in lengths from 8 to 12 feet.

Pre-decorated gypsum board This type of gypsum board has a decorative surface not requiring further treatment. The surfaces may be coated, printed, or have a vinyl film. Other pre-decorated finishes include factory-painted and various textured patterns. Pre-decorated panels are also available with type-X core.

Moisture resistant (MR) gypsum board Moisture-resistant (MR) board is a specially processed gypsum wallboard for use as a base for ceramic tile and other non-absorbent finish materials in wet areas (Fig. 9-5). The core, face paper, and back paper of MR board are treated to withstand the effects of moisture and high humidity. MR board is available in the standard 4-foot width and 8 foot and 12 foot lengths. Its facing paper is colored green (hence, often called "green board" by drywall mechanics) so as to make it readily distinguishable from regular gypsum wallboard. It is also available in ½-inch regular, and ½-inch or ⅝-inch type-X core.

Foil-backed gypsum board These boards are made by laminating special kraft-back aluminum foil to the back surface of regular or X-type panels. They form an effective vapor retarder for walls and ceilings when applied with foil surfaces next to the framing on the interior side of an exterior wall in single-layer applications or as the base layer in a multi-layer system.

9-5 *Moisture-resistant (MR) or water-resistant (WR) board can be used as a base for ceramic tile.*

Foil-baçked gypsum panels provide a water vapor retarder to help prevent interior moisture from entering wall and ceiling spaces. The permeability of the total exterior wall is dependent on the closure of leaks with sealants at the periphery and penetrations such as outlet boxes.

Foil-backed gypsum panels are designed for use with furred masonry, wood or steel framing. These panels are generally available in ⅜-, ½-, and ⅝-inch thicknesses. Foil-backed gypsum board is also available with an X-type core.

High-strength ceiling board Some gypsum board manufacturers provide a ½-inch high strength ceiling board as a cost-saving alternative to ⅝-inch wallboard for ceilings. High strength ceiling board has a specially formulated core that resists the natural tendency to sag when attached to ceiling joists and trusses spaced 24 inches O.C. and where heavy water-based textures are applied.

Gypsum ceiling panels Gypsum ceiling panels have a noncombustible gypsum core with pre-decorated surfaces of an attractive appearance and high light reflectance. The panels install easily in standard exposed grid systems. Because ceiling panels have a rigid gypsum core, they resist sagging and warping and do not require clips to offset bowing. Most gypsum ceiling panels are accepted by the United States Department of Agriculture (USDA) for use in food-service and food-processing areas. Most gypsum ceiling panels are ½ inch thick with a 2-mil white stipple-textured vinyl laminate for high light reflective properties and easy cleaning, and are available in 2-×-2 foot and 2-×-4 foot panels for interior and unexposed exterior applications. In accordance with industry practice, these dimensions are nominal. Actual sizes, correspondingly, are

$$2 \times 2 \text{ feet} = 23\tfrac{3}{4} \times 23\tfrac{3}{4} \text{ inches}$$
$$2 \times 4 \text{ feet} = 23\tfrac{3}{4} \times 47\tfrac{3}{4} \text{ inches}$$

Custom sizes are available from most manufacturers by special order.

Gypsum shaft liner Gypsum shaft liner is used for the construction of elevator, air and mechanical shafts, stairwell walls, and general partitions. It is designed for use on building construction that requires a two-hour fire-resistance rating. An arrangement constructed with non-load-bearing metal framing, a 1-inch type-X shaft liner panel, and two layers of regular ½-inch gypsum boards will help to protect the framing members and prevent or delay the spread of fire from one area to another. With the addition of insulation or resilient channels, the construction will effectively reduce sound transmission. Liner board is available in ¾- or ½-inch thicknesses in widths of 24 or 48 inches, and with square edges (sometimes eased square edges). Shaft liners are designed to be non-load-bearing.

Gypsum sound-deadening board When used in conjunction with either ½- or ⅝-inch type-X board a ¼-inch sound-deadening board will provide an excellent sound- and fire-resistant installation. The ¼-inch sound-deadening board is noncombustible, easy to apply, and adds structural integrity. In U_L designs, one layer of ½-inch type-X panel, with one layer of ¼-inch sound-deadening board on each side of wood or metal studs, will provide one hour of fire protection and 45 to 50 STC sound isolation.

Gypsum base for veneer Gypsum base for veneer plaster is used as a base for thin coats of durable, high strength, wear-resistant gypsum veneer plaster. The gypsum boards come in 4-foot widths with square cut edges. Standard veneer-base drywall is ⅜-, ½- or ⅝-inch thick. Type-X is only available in widths of ½, or ⅝ inch. Both types come in 8-, 10-, and 12-foot lengths. The gypsum core is faced with specially treated, multi-layered paper designed to provide a maximum bond to veneer plaster finishes. The paper's absorbent outer layers quickly and uniformly draw moisture from the veneer plaster finish for proper application and finishing; the moisture-resistant inner layers keep the core dry and rigid to resist sagging.

Solid wood paneling Red oak, pecan, walnut, cherry, knotty pine, redwood, and white oak are among the types of woods that have been and are still being used as interior wall coverings. Whether stained, varnished, or natural finished, solid wood paneling is extremely susceptible to moisture damage. For this reason, be sure not to store or install the product where moisture absorption is likely to occur.

Many varieties of solid wood paneling are intended for application in pieces of random widths from 4 to 8 inches in order to emphasize the natural solid wood appearance. Nominally 1 inch thick, solid wood paneling ranges in length from 3 to 10 feet:

- *Plywood.* A less expensive way of portraying the natural wood look in an area is with plywood sheets (or panels) with outer plies of wood veneer. The variety of face plies available is vast and can resemble plank walls with enough versatility to be used in almost any interior area of the structure. ¼-inch plywood sheets are the most commonly used for finishing interior walls; however, ⅜- and ¾-inch panels are available. Although special sizes are available for jobs requiring them, the normal plywood panel is 4 feet wide and 8 feet long.
- *Hardboard.* Composed of wood fibers placed under pressure and thermally treated for bonding purposes, hardboard is generally manufactured in sheets 4 feet wide by 8 feet long and ¼ inch in thickness. Other sizes are

available for specific needs. The finish surface of the panels can resemble a variety of wood with a smooth or scored texture or can be manufactured to imitate marble, tile or painted surfaces.

- *Tile.* Although ceramic tile is applied over a sub-finish (such as moisture resistant gypsum board) instead of directly to the framing of the structure, it is a widely used product, particularly in baths and kitchens. An enormous variety of tile is available, ranging in sizes from 1 inch square to 12 inches square, tiles coming in just about any color and shape imaginable. Used in conjunction with the appropriate adhesive, tile provides a low maintenance, long-wearing, water-resistant surface.

Determining required quantities

The thickness of a dry-wall finish will be determined (at least in part) by the framing's spacing on center. Table 9-1 presents the generally accepted thicknesses of plywood, hardboard, gypsum board, and wood paneling as determined for 16, 20 and 24 inch on center spacing.

For solid wood paneling in the form of individual boards, estimating material quantities is the same as for board subflooring. Multiplying the length and height of each wall, totalling these figures, and subtracting a figure equal to the area which doors and windows will occupy in the room will result in the measure of the area to be covered, to which 15 percent should be added as a waste factor. For example, the calcula-

Table 9-1 Recommended minimum thickness for interior dry-wall products.

Spacing of framing inches on center	Plywood and hardboard	Wood paneling	Gypsum board
16 in. O.C.	¼ inch	⅜ inch	⅜ inch
20 in. O.C.	¼ inch	½ inch	½ inch
24 in. O.C.	5⁄16 or ⅜ inch	25⁄32 inch	½ inch

tions for a room measuring 12 × 20 × 8 with openings totaling 48 square feet would appear as follows:

$$2(12 \times 8) + 2(20 \times 8) = 512 \text{ square feet}$$

$$512 - 48 = 464 \text{ square feet}$$

$$464 \times 0.15 = 69.6 \text{ square feet}$$

$$464 + 69.6 = 533.6 \text{ square feet}$$

Rounded to standard sizes, the final figure becomes 533½ square feet of area to be covered. Again, as with sub-flooring, solid wood paneling is generally sized as nominal 1 inch in thickness; consequently, the number of square feet is equal to the number of board feet.

The same calculations can be used when estimating the area to be covered for any finishing product which is manufactured in 4-×-8 sheets, such as plywood, hardboard, or gypsum-board. Once the gross area of coverage has been determined, wall openings for windows and doors can be accounted for by deducting a half panel per door and a quarter panel per window. After the appropriate deductions have been made and a 15 percent waste factor added, the resulting total area to be covered is divided by the area of one panel (4 × 8 = 32 square feet).

In order to calculate the number of tiles necessary to cover a space, the area of that space must again be figured. The number of tiles contained within one square foot of area needs to be determined next. For example, a standard tile 4½ inches square will cover approximately 18 square inches (4½ × 4½ = 18.06). One square foot equals 144 square inches, and 144 divided by 18 = 8. Since eight standard 4½-inch tiles will cover one square foot, multiplying the total square footage of the area to be covered by 8 will yield the number of tiles needed for the job.

Finish flooring & coverings

The assortment of materials used as finish flooring is expansive: polished marble, hardwoods, softwoods, different forms of tiles, carpeting, and vinyl (plus more!). Whichever type of finish flooring is being installed, it was probably selected for one or more characteristics unique to that product. Comfort, durability, maintenance, appearance, and cost are some of the most scrutinized factors a buyer considers prior to specifying the material to be used.

Types of finish flooring

- *Hardwood flooring.* In recent years, the popularity of fine quality hardwood floors has risen. In general, this type of wood-strip finish floor comes in random lengths and widths to break up any semblance of pattern that would infer anything other than individually laid pieces of wood. The finish flooring is usually of tongue and groove design with an installation often requiring furring strips spaced 12 or 16 inches apart, secured to the subflooring.
- *Vinyl.* Vinyl flooring is normally manufactured in either rolls or tiles. Rolls of 6 feet, 9 feet or 12 feet in width with varying lengths are available, while tiles usually come 9 inches or 12 inches square. This resilient material has a very low maintenance factor and is durable enough to last for years even in heavy traffic areas. Available in a multitude of colors and styles, vinyl can be used to accent the interior of a structure by imitating such natural flooring as brick, marble, or slate.
- *Carpeting.* Carpeting manufactured of natural fibers such as wool have been joined by a wide array of man-made fiber carpets. Together, the natural and synthetic carpets provide the potential to cover any flooring area with carpeting. For years now, kitchens and baths have utilized synthetic fiber carpeting to provide the comfort, sound-insulating, and non-slip safety factors that were once the province of living rooms and formal areas only.

Like roll vinyl, carpeting usually comes in rolls of varying widths and lengths. However different-sized squares of carpeting are available, particularly in varieties of product intended for heavy use (such as indoor/outdoor). Colors and patterns are practically limitless, allowing carpeting to fit into just about any motif.

Calculating quantities of finish flooring

Determining the amount of material necessary for carpeting or vinyl usually involves measuring for roll goods. Making accurate measurements is simply a matter of multiplying the length times the width of the area(s) to be covered, which will provide the square foot area of the room(s). Because both of these products are normally discussed in terms of square yards, dividing the total square footage by 9 will convert square feet into square yards.

Now that the total number of square yards is known, do not forget to add a waste factor; 15 or 20 percent would be acceptable for these products. Material for each area to be covered should be ordered separately, providing roll material of a width matching that of the individual floor space.

Carpet runners for stairs are sized by the linear yard in widths ranging from 18 inches to 36 inches. To compute the linear yardage of a stairway, follow these steps:

1. Measure the depth (in inches) of a single tread: 7¼ inches.
2. Measure the height (in inches) of a single riser: 10¼ inches.
3. Combine the number of inches from Steps #1 and #2: 17½ inches.
4. Multiply the sum in Step #3 by the number of stairs: 17½ × 14 = 245.
5. Add the length (in inches) of any landings to this figure: one 3-foot landing: 245 + 36 = 281.
6. Divide the sum in Step #5 by 36 to convert to linear yards: 281 / 36 = 7.8 (round to 8 linear yards for this particular stairway).

Tiles

Since vinyl tiles and carpet tiles share some common sizes, Table 9-2 has been designed for use in comparing the number of tiles required for a given number of square feet depending

Table 9-2 Comparison of area covered by finish flooring tiles of different sizes.

Square foot area to be covered	Number of 9 × 9 tiles needed	Number of 12 × 12 tiles needed	Number of 9 × 18 tiles needed	Square foot area to be covered	Number of 9 × 9 tiles needed	Number of 12 × 12 tiles needed	Number of 9 × 18 tiles needed
1	2	1	1	60	107	60	54
2	4	2	2	70	125	70	63
3	6	3	3	80	143	80	72
4	8	4	4	90	160	90	80
5	9	5	5	100	178	100	90
6	11	6	6	200	356	200	178
7	13	7	7	300	534	300	267
8	15	8	8	400	712	400	356
9	16	9	8	500	890	500	445
10	18	10	9	600	1,068	600	534
20	36	20	18	700	1,246	700	623
30	54	30	27	800	1,424	800	712
40	72	40	36	900	1,602	900	801
50	89	50	45	1,000	1,780	1,000	890

upon the size of the tiles used. Again, the calculations are based on the total square feet of area to be covered. Once known, the square footage is divided by the area occupied by one tile. For example, 1 square foot is equal to 144 square inches; a 9-inch by 9-inch tile equals 81 square inches. It would take two of the 9-×-9 tiles to cover 1 square foot, but it would only take one 12-×-12 tile to cover the same area.

Wood-strip finish flooring

In determining the amount of flooring material required for a given space, the square footage of the area to be covered must be calculated. As with other wood products, finish wood-strip flooring is generally nominal 1 inch (25/32); consequently the square foot figure for the space also equals the number of board feet needed for that space.

The amount of additional material which should be allocated for waste purposes ranges from 50 percent to 25 percent depending on several factors: The experience of the individual crew installing the floor, the size of the flooring, and the number and size of any projections (such as bay windows) present in the space. Table 9-3 provides suggested guidelines for waste factors depending on the size of the wood-strip flooring.

Interior mill work

Much of what is considered interior mill work constitutes those details without which a space would not appear finished.

Table 9-3 Suggested waste factors for use with various size wood-strip flooring.

Size of wood flooring strips (inches)	Percentage waste factor
½ × 2 (1 × 21½ nominal)	25
½ × 1½ (1 × 2 nominal)	33⅓
⅜ × 2 (1 × 2½ nominal)	25
⅜ × 1½ (1 × 2 nominal)	33⅓
25/3 × 3¼ (1 × 4 nominal)	25
25/32 × 2¼ (1 × 3 nominal)	33
25/3 × 1½ (1 × 2¼ nominal)	50

Types of mill work

Whether referring to the casings around doors and windows, chair rails, baseboards, wainscot caps, cornices, half rounds, quarter rounds, balusters, or the many different moldings which are available; these are all types of mill work.

Calculating quantities

With the exception of special order trim, which will sometimes be quoted in terms of pieces, most all mill work will be discussed in terms of linear feet. Determining the amount of baseboard needed for a 10-×-12 room involves calculating the perimeter of the room (adding together the lengths of all 4 walls). For the example above, 2 (10) + 2 (12) or 10 + 10 + 12 + 12 = 44 linear feet.

Windows & doors

The variety of doors and windows today is almost mind-boggling; selecting the types of doors for specific purposes is no longer an automatic decision. The choices between wood or metal doors, hollow core, or solid core just scratch the surface. The realm of windows provides even more choices; wood, aluminum or vinyl clad, double hung, slider, single pane, double pane, or even triple pane.

Regardless of what type, or combination of types of doors and windows have been stipulated for the structure, each one will normally be priced out as a separate unit. The sizing of each unit is equal to the rough openings left in the framing to accommodate them. The concern of the contractor is in estimating the labor costs associated with installing these items. Once again, labor hours are equal to the number of hours required to complete the task multiplied by the number of laborers performing the work. For example, two laborers requiring 1 hour to install a double hung window equals 2 labor hours. Multiplying the number of labor hours times the wage of the laborers will yield the labor cost of installing that window.

Interior painting

Most interior paints are latex base and dry to either a flat or semi-gloss finish.

Estimating quantities

Like exterior paints, interior paints have average coverage areas printed on their labels. These figures are averages, so depending on the type of surface to be covered this average may vary. By dividing the average coverage area into the actual area to be covered, the quantity of paint required can be determined. Remember, if more than one coat of paint is to be applied, the quantity must be increased accordingly (see Table 7-3 in Chapter 7).

Estimating labor costs

The average time an experienced painter requires to cover various interior surfaces is presented in Table 9-4. These figures are based on surface areas of 100 square feet for areas such as walls and ceilings, and 100 linear feet for trim areas. Windows and doors are estimated per unit.

Table 9-4 Approximate labor hours per task for interior painting.

Surface and procedure	Unit of measure	Average time required
Wallboard or plaster		
Sealer	100 sq. ft.	25 to 30 minutes
Oil-based paint	100 sq. ft.	20 to 25 minutes
Latex-based paint	100 sq. ft.	10 to 20 minutes
Trim	100 linear feet	
Sanding	"	45 to 60 minutes
Varnishing	"	30 to 45 minutes
Enameling	"	30 to 45 minutes
Painting	"	30 to 45 minutes
Staining	"	25 to 35 minutes
Doors and windows	Per unit	
Paint		30 to 45 minutes
Floors	100 sq. ft.	
Power sanding	"	45 to 60 minutes
Filling	"	30 to 40 minutes
Shellacking	"	15 to 20 minutes
Varnishing	"	25 to 30 minutes
Waxing	"	25 to 30 minutes
Machine polishing	"	20 to 30 minutes

Wallpaper

Aside from paint, wallpaper is probably the most often used product for interior wall covering. Available in a wide range of colors, prints, and patterns, many different textures are also available.

Determining materials required

Wallpaper is normally supplied by the roll with about 36 square feet of material on each roll. Allowing for waste each roll is reduced to approximately 30 square feet per roll. Boarder strips, used for trimming and matching, are generally supplied by the linear foot.

In determining the quantity of wallpaper for a project, determine the area of the space to be covered. Divide this figure by 30 to obtain the gross number of rolls required for the job. From this number, subtract one roll for every two openings (windows or doors) to arrive at the net number of rolls required.

Review questions

1. You need to build a wall using 2×4s. The wall is to be 20 feet long and 8 feet high. How many 2×4s are required if the 2×4s are on 16" centers?
2. How many sheets of sheet rock will be needed for the example in question 1?
3. A job in Chicago, Illinois, requires an order of ponderosa pine board items. The order consists of 1" × 6" boards 8' long, 10' long, 12' long, 14' long, and 16' long. The cost of the product at the mill is $530 per thousand board feet. One board foot of lumber equals a 1" × 12" board 1 foot long.

   Freight cost for shipping a carload of lumber to Chicago from the mill at Kettle Falls, Washington, is $4420. The amount of lumber that ships on a carload is 98,000 total board feet of lumber.

   What is the cost of the individual boards in the order.
4. A family room measures 18′ × 24′. The cost of carpet is

$13.00/square yard and the cost to install the carpet is $5.00/square yard. What is the cost for carpet, including installation?

5. If paint covers 300 square feet per gallon, how many gallons will it take to paint a home with a flat roof measuring 47 feet wide and 10 feet high?

10

Math for heating & cooling

HVAC systems and equipment are among the most significant components determining habitability and energy consumption in buildings today. Designers have significant latitude in the selection and installation of HVAC systems. A poorly designed system can easily have twice the yearly energy costs of an energy conserving design, while not keeping the structure comfortable.

System interactions complicate analyzing the energy use and cost of an HVAC system. An efficient system is not just one that uses efficient equipment. System level efficiency takes into account installation, control, maintenance, system losses, and component interactions (such as reheat or heat recovery). An efficient system will minimize energy use by minimizing system losses, maximizing equipment efficiencies, utilizing free heating/cooling, and recovering heat where possible.

In addition, good operating and maintenance techniques are essential to achieve the energy efficiency contemplated in the design. The following fundamental factors are key to improving HVAC system efficiency:

- Reducing system losses from ductwork and piping;
- Reducing system operation through the use of automatic time controls and zone isolation;
- Reducing system operation through requirements for zonal controls;
- Reducing system inefficiencies by limiting equipment oversizing;

- Reducing distribution losses, limiting HVAC fan energy demand, and requiring efficient balancing practices.

Although compliance with local codes assures a minimum design level of HVAC system and equipment performance, designers should be encouraged to view the requirements as a starting point and investigate designs that exceed these minimums. For example, application of heat recovery or high efficiency equipment can create a system that is more efficient than the code requires while exhibiting an excellent return on investment.

Load calculations

The designer must make heating and cooling load calculations before selecting or sizing HVAC equipment. The purpose of this requirement is to ensure that equipment is neither oversized nor undersized for the intended application.

Oversized equipment not only increases owner costs, but usually operates less efficiently than properly sized equipment. It can also result in reduced comfort control due to such factors as lack of humidity control in cooling systems and fluctuating temperatures from short-cycling. Undersizing will obviously result in poor temperature control in extreme weather. While load calculations are required, there are generally no requirements that actual equipment sizes correspond to the calculated loads.

Accurate calculation of expected cooling loads begins with a reliable calculation methodology. Many code authorities require that calculation procedures comply with the fundamentals set out by the American Society of Heating, Refrigeration, and Air-Conditioning Engineers (ASHRAE) or a similar computational procedure. This is not to preclude the use of other time-proven methodologies that may not precisely follow ASHRAE procedures, such as those developed by some major equipment manufacturers and other professional groups.

At this time, there is no universal agreement among engineers on a single load calculation procedure. Because the thermodynamic performance of buildings and HVAC systems is so complicated, the variety of available procedures produce results that vary considerably. For this reason all calculation methods and computer software must have simplifying assumptions embedded within them to make them practical to use. Depending on the application, these simplifications can result in inaccuracies and errors. Designers need to be aware of the limitations of the calculation tool used and apply "reality checks" to the results, which are based on past "real life" experience, to avoid sizing errors.

Once a reliable calculation procedure is selected, accurate load calculations then depend on the use of accurate design parameters. Many local codes restrict some of these parameters, but also allow for flexibility and judgment on the part of the designer.

The comfort envelope

The perception of thermal comfort is a function of many variables including air temperature, radiant temperature, air movement, gender, clothing, activity level, age, and adaptation to local climate. Therefore, there is no single design temperature or humidity recommended by ASHRAE to achieve acceptable comfort.

The summer comfort envelope extends to operative temperatures as high as 78 to 80°F, depending on the humidity. However, practical experience and recent research suggests that the comfort range in commercial applications may actually be much lower, with its outer border being near the outer border of the winter comfort band (around 73 to 76°F depending on relative humidity). Designers should be aware of this if selecting higher design cooling temperatures for comfort HVAC applications since, regardless of energy consequences, the ultimate purpose of the HVAC system is to maintain comfort.

The comfort envelope applies only to sedentary people wearing typical clothing. It may be necessary to adjust conditions to expected clothing levels in the space (e.g., doctors' examination rooms), occupant age (e.g., retirement homes), and activity level (e.g., health clubs).

Envelope

Building envelope performance parameters, such as thermal transmittance (U-values), shading coefficients of fenestration, wall and glazing areas, etc., must be consistent. For instance, if a low transmission glass is stipulated for one or more areas of the structure, the same glass must also be assumed for HVAC load calculations.

Take care to account accurately for the potentially significant effect of wall and window framing systems on thermal conductance to avoid seriously underestimating loads. Thermal bridging is particularly significant for metal framing, which can reduce the insulating effect of wall insulation or double glazing by 50% or more. See Chapter 7 for further information.

Other loads

Internal cooling loads due to people, equipment such as computers and copiers, and lighting are often difficult to predetermine because they vary so much from one building to another. Most codes leave the designer a great deal of flexibility in determining design parameters for these miscellaneous loads, normally requiring only that they be compiled from one or more of the following sources:

- Actual information based on the intended use of the building. For instance, the number of seats can determine the exact occupancy of a theater.
- Published data from manufacturers' technical publications. This is most useful for large data processing equipment for which data is readily available.

NOTE: Take care to use the data correctly; in most cases equipment manufacturers list peak operating power, however, most equipment will usually only operate for a relatively short period of time at peak power. For instance, a copy machine will use much less power when idling than when copying. A laser printer will operate at peak power only a few seconds each hour when its toner heater operates. The designer should apply an appropriate load diversity factor to such equipment when it is known to operate intermittently. Other manufacturers' equipment ratings are similarly misleading, particularly those for popular personal computers and peripherals. This can lead to significant cooling system oversizing.

Once total heating and cooling loads have been determined from the data and procedures previously outlined, they may be increased by as much as 10%. This increase allows for additional future loads or unexpected loads that may arise as space usage changes.

Pick-up loads

Load calculation methods need to take into account some of the effects of thermal mass as it affects cooling loads. For instance, solar heat gain factors vary as a function of space mass. This accounts for the delay that occurs from the time the floor of the room and the furnishings absorbs solar radiation until that radiation is released as heat into the room air. At the time of release into the room air, the radiation becomes a component in the air conditioning load.

Similar factors are applied to internal gains such as lights and people. However, these methods generally assume that the space is maintained at relatively constant temperatures, and thus the loads they predict often are referred to as steady-state loads.

When HVAC systems are operated intermittently, the system must be capable of warming up or cooling down spaces whose temperatures have been left to float out of comfortable ranges. These are called pick-up loads. There are two generally accepted methods of determining these loads.

- Calculate from basic principles, based on the heat capacity of the space, the setback or setup temperatures and the amount of time allowed to bring the space to operating temperatures.

While this method would seem to be capable of producing the most accurate answers, in practice, it is seldom used because the heat transfer modes are so complicated. For instance, the internal walls and floors of a building, and much of its furnishings, seldom reach a steady state during setback. Although the air in the space may have risen or dropped in temperature significantly during HVAC system shutdown, most of the building mass has not. The building mass lags behind due to its heat capacity and the limited heat transfer avenues to the air within the space. In pick-up load calculations, assuming that the entire mass of a building was cooled or heated to the setback/setup temperature would result in grossly oversized HVAC equipment. The non-steady-state heat transfer that occurs as spaces warm-up or cool-down complicates calculations even further.

- The most common method of calculating pick-up loads is to apply warm-up or cool-down factors to the steady-state loads calculated by conventional methods.

Most codes allow steady-state heating loads to be increased by as much as 30% and steady-state cooling loads to be increased by as much as 10% to allow for pick-up loads. The starting steady-state loads may include the 10% safety factor.

The heating load warm-up factor is larger than the cool-down factor because warm-up generally occurs in the morning at the same time that peak heating loads occur. On the other hand, cool-down loads generally occur before solar loads and many internal heat gains are present, so the cool-down factor need not be as large.

Zones

An HVAC thermostatic control zone is defined as a space or group of spaces whose load characteristics are sufficiently similar that the desired space conditions can be maintained throughout with a single controlling device. Normally, codes require that the cooling to each zone be controlled by an individual temperature controller sensing the temperature within the zone. To meet this requirement, spaces must be grouped into proper control zones. For instance, spaces with exterior wall and glass exposures must not be zoned with interior spaces. Similarly, spaces with windows facing one direction should not be zoned with windows facing another orientation unless the spaces are sufficiently open to one another that air may mix well between them to maintain uniform temperatures.

In many commercial, industrial, and (today, more frequently) residential situations, spaces housing temperature- or humidity-sensitive equipment or processes are located adjacent to spaces that need only comfort conditioning. To avoid the waste of energy due to overconditioning the non-process spaces, smart designers serve these areas with separate air handling systems: one controlled for comfort purposes and the other controlled as required by the process. Alternatively, the two spaces may be served by a single system if it is controlled only as required for comfort, while supplementary equipment (such as humidifiers and auxiliary cooling equipment) is added to maintain the process requirements.

Generally, separate air distribution systems are not necessary if the spaces requiring comfort-conditioning use no more than 25% of the total system air supply or if they do not exceed 1000 square feet. It is recommended (although not normally required) that zones with substantially differing load characteristics be served by separate air distribution systems.

For instance, interior spaces, which have a relatively constant requirement for cooling regardless of weather conditions, should be served by systems separate from those serving perimeter spaces whose loads are more dependent on solar loads. Zonal systems, such as single-zone packaged units, fan-coils, hydronic water-cooled packaged units, and room air conditioners meet this requirement inherently since each zone is served by a separate system. However, central multiple zone systems such as variable air volume (VAV) systems are often designed to serve many zones that may have widely varying load characteristics. This can lead to some inefficiencies.

Another inefficiency may arise in interior zones. To keep perimeter zone supply volumes reasonable at peak cooling conditions, supply air temperatures may generally be in the range of 50–60°F. This may result in supply quantities that are too low for good circulation in interior spaces; particularly in those designed with energy efficient lighting. As a solution, many designers install fan-powered boxes in interior spaces that mix plenum air with supply air to keep overall supply volumes up. Requiring air to be overcooled wastes energy. The overcooled air is reheated with plenum air that is warmed by radiant heat given off from recessed light fixtures. Also, the fan-powered box fans and motors are typically inefficient, often requiring more power than even a central fan despite the higher duct losses typical of central systems.

In general, these inefficiencies result in only relatively small energy losses, so in most applications installing separate systems for interior and perimeter zones is not cost effective.

Exception—Two independent systems

(a) This exception applies to zones, typically perimeter offices that are served by two independent HVAC systems. One of the two systems, called the perimeter system, is designed to offset only skin loads, those loads that result from energy transfer through the building envelope. Interior loads, such as those from lights and people, are controlled by a second system called the interior system. The exception is permitted only if: the perimeter system has at least one zone for each major exposure, defined as an exterior wall that faces 50 contiguous feet or more in one direction.

For example, in Fig. 10-1, a zone must be provided for each of the exposures that exceeds 50 feet in length, while the shorter exposures on the serrated side of the building need not have individual zones.

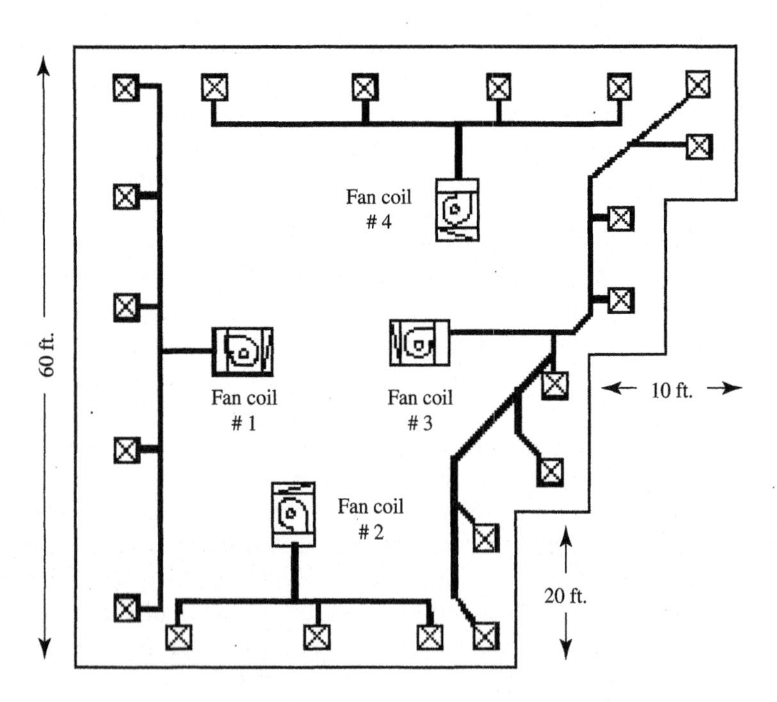

10-1 *A zone must be provided for each exposure that exceeds 50 feet in length.*

One, or more, thermostats located in the zones served controls each perimeter system zone. In Fig. 10-1, this requirement might be met by controlling the perimeter fan-coil with one of the thermostats controlling one of the four interior systems VAV zones on the exposure. Alternatively, all four thermostat signals could be monitored and the one requiring the most cooling be used to control the fan-coil. Finally, a completely independent thermostat could be installed in one of the rooms on the exposure to control the fan-coil, set to a setpoint that was below those controlling the VAV boxes.

Exception—Dwelling units

(b) A dwelling unit, such as an apartment or condominium may be considered a single control zone. Bedrooms and living areas are best served when zoned separately to allow for different operating temperatures during the day and night. However, code specifications usually do not require rooms of an apartment to be zoned by exposure or by function.

Setpoint requirements

In general, zone thermostats used to control space cooling must be capable of being set up to 85°F or higher. Thermostatic controls may be set either locally (with adjustment buttons, switches, knobs, etc.) or remotely, such as by direct digital controls (DDC). Replacing elements for thermostats may also change setpoints when the setpoint is a function of the sensing element.

Off-hour time controls

Most HVAC systems serve spaces that are occupied on an intermittent basis, but in a fairly predictable manner. To reduce HVAC system energy usage during off-hours, local codes are changing to require HVAC systems be equipped with automatic controls that will shut off the system or set back setpoints. Examples of acceptable automatic controls are time clocks, programmable time switches, energy management systems, direct digital control systems, wind-up bypass timers, and occupancy sensors.

Since the term "HVAC system" applies to all equipment that provides any or all of the ventilation or cooling functions, requirements usually include time controls on systems ranging from simple ventilation fans to large chiller plants. There are, however, three exceptions where time controls are not required:

- Systems serving spaces that are expected to be in continuous operation. Examples include hospitals, police stations and detention facilities, central computer rooms, and some 24-hour retail establishments.
- Where it can be shown that setback or shutdown will not result in an overall reduction of building energy costs. There are few cases where this might occur.
- Small equipment, those with full-load demands of 2 kW (6826 Btu/h) or less, may have readily accessible manual on/off controls in lieu of automatic controls. This exception is intended to apply to small independent systems such as conference room exhaust fans or small toilet room exhaust fans.

The intent is that all energy associated with the operation of the equipment be included in the 2 kW. For instance, a fan-coil that would use chilled or hot water, requiring operation of a remote chiller or boiler, may use in excess of 2 kW of energy when it operates. In this case, the fan-coil would have to be automatically controlled, although many fan-coils may be interlocked to the same time clock.

Off-hour isolation

Large central systems often serve zones that are occupied by different tenants and may be occupied at different times. When only a part of the building served by the system is occupied, energy is wasted if unoccupied spaces are also conditioned.

To minimize this waste, energy conservation requirements stipulate that systems serving zones that can be expected to operate non-simultaneously for 750 hours or more per year be equipped with isolation devices and controls that allow each zone to be shut off or set back individually. Normally, zones may be grouped into a single isolation area, provided that:

1. The total conditioned floor area of the group does not exceed 25,000 square feet.
2. All zones in the group are located on the same floor.
3. Spaces expected to be unoccupied only when all other spaces are unoccupied need not be isolated. For example, isolation would not be required for the entry lobby of a multipurpose building since it is occupied when any of the building areas are in operation. This lobby would not benefit from isolation since it would need conditioning whenever the HVAC system is on.

In many cases, the eventual occupants of the building are unknown when the HVAC system is designed, such as with speculative buildings. In that case, isolation zones may be predesignated provided they do not violate either the 25,000-square-foot or one-floor rule. If occupant schedules are unknown, assume that isolation will be required and make appropriate provisions in the HVAC system design.

Each isolation area must include individual automatic time controls as if it were a separate HVAC system. This will allow each isolation zone to automatically operate on different time schedules. Figure 10-2 shows a schematic riser diagram of a central VAV fan system serving several floors of a building, each assumed to be less than 25,000 square feet. Isolation of each floor is required if they are to be occupied by tenants that can be expected to operate on different schedules, or if tenant schedules are unknown.

Any one of the methods depicted schematically in Fig. 10-2 may easily accomplish isolation of floors or zones:

- On the lowest floor, individual zones are controlled by direct digital controls. If the direct digital control software can be programmed with a separate occupancy time schedule for each zone or for a block of zones, isolation can be achieved without any additional hardware. The boxes are simply programmed to shut off or control setback setpoints during unoccupied periods.

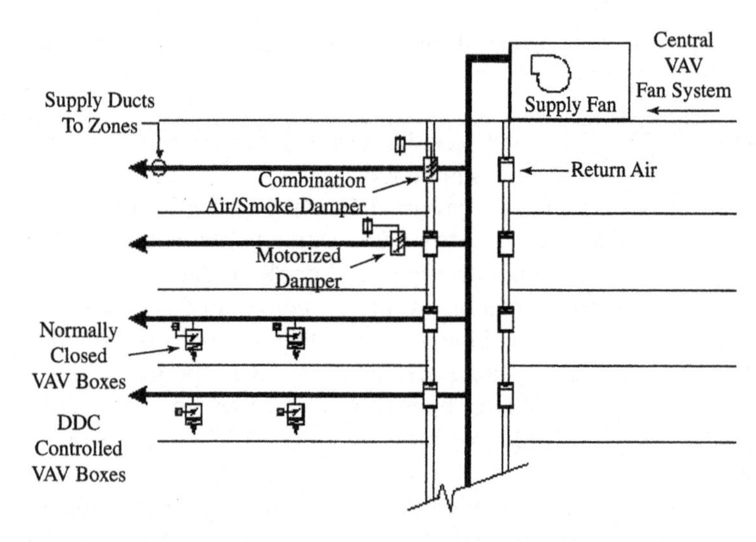

10-2 *Schematic riser diagram of a central VAV fan system serving several floors*

- On the next floor up, zone boxes are shown to be "normally closed," which means when control air or control power is removed, a spring in the box actuator causes the box damper to close. This feature can be used as an inexpensive means to isolate individual tenants or floors. The control source to each group of boxes is switched separately from other zones. When the space is unoccupied, the control source is shut off, automatically shutting off zone boxes. A separate sensor in the space can restore control to maintain setback or setup temperatures.
- On the next floor up in Fig. 10-2, isolation is achieved by simply inserting a motorized damper in the supply duct. On the top floor, the cost of this damper is saved or reduced by using a combination fire/smoke damper at the shaft wall penetration. Smoke dampers are often required by life-safety codes to control floor air flow for pressurization. These dampers may serve as isolation devices at virtually no extra cost, provided they are wired so that life-safety controls take precedence over off-hour controls. (Local fire officials generally allow this dual usage of smoke dampers

and often encourage it since it increases the likelihood that the dampers will be in good working order when a real life-safety emergency occurs.)

Note that on all floors in Fig. 10-2, shutoff is not shown on floor return openings. This represents the common requirement that only "the supply of cooling" be shut off. In addition, with a plenum return system, the amount of air drawn off an unconditioned floor will be negligible compared to the occupied floors that have positive air supply, since the latter will be pressurized.

In order to effect proper zone control in many spaces, a positive means of zone shutoff or setback is required. Shutoff VAV boxes (boxes with no minimum volume setting) cannot be assumed to automatically close during unoccupied periods due to low loads. 24-hour internal loads (such as PCs, idling copy machines, or emergency lighting) can combine with continuous envelope loads, which would prevent shutoff or setback.

Simply providing a means for central system zone isolation does not end the design task. Central equipment must be capable of operating at the very low loads that can be expected when only one isolation zone is operating. Experience has shown that almost any fan with a variable speed drive for static pressure control can operate with stableness to zero flow.

Large axial fans with variable pitch blades may also be able to operate at low flows without overpressurizing ductwork. Fan curves at minimum blade pitch should be reviewed to be sure shutoff pressures are below duct design pressures. Where fans cannot be selected to operate safely at low loads, large fans can be broken into smaller fans in parallel with operation, staged so only one fan operates at low loads.

The same considerations can apply to central chiller plants. The plant must be able to operate at low loads for extended periods. If frequent chiller cycling is not acceptable, either multiple or staged chillers can be used. As a last resort, hot-gas-bypass can be used to maintain stable low-load operation, but this can significantly increase energy costs and should only be used if installation cost budgets prohibit the use of multiple staged chillers.

Ventilation effectiveness

Outdoor air for ventilation must be effectively mixed with air in the occupied space. Poorly selected, sized, or placed air outlets can reduce ventilation effectiveness to as little as 50% (with 100% defined as perfect mixing), although this is unusual. To mitigate this problem, outdoor air intake rates are often increased, or overall circulation rates are increased (by using fan-powered mixing boxes or smaller supply air to room air temperature differences), to improve air outlet performance. Both of these options will increase energy usage unnecessarily. Ventilation effectiveness can be improved with minimal energy impact by following these guidelines:

- Use supply air outlets that have high aspiration ratios, such as slot diffusers or light troffer diffusers. The air pattern a few feet from a well-designed outlet supplied with a small amount of cold (or hot) air can be identical to the pattern that results from a poor outlet supplied with a higher air quantity. In other words, ventilation effectiveness can be improved without arbitrarily increasing supply air quantities, which unnecessarily increases fan load.
- Distribute outlets well around each space; avoid using one large outlet when several small outlets will distribute air more evenly.
- Do not oversize outlets, which reduces their throw and aspiration ratio. This is particularly important for VAV systems, which will operate at less than full flow most of the time.
- Locate returns where they will not short-circuit supply air. With properly sized outlets, the location of the return will generally not affect space mixing unless the return is located too close to the supply.
- Take extra care when using light fixtures for returns since they are often close to supplies and their location is not under the control of the HVAC designer. Ensure that fixtures located close to supplies are blanked off.

Intermittently occupied spaces

For large spaces that have high peak occupancies but are only intermittently occupied, such as ballrooms, meeting rooms, and theaters, outside air intake should be varied rather than constantly maintaining the high rates needed for peak occupancy. The ASHRAE Ventilation Rate Procedure allows for peak ventilation rate reduction for spaces occupied for only short periods. Where peak occupancy occurs for less than 3 hours, determine ventilation rates by the average occupancy or one-half the peak occupancy, whichever is larger, over the duration of system operation.

These large spaces that are people-load dominated, can benefit from use of a VAV supply system. VAV supply systems vary supply air with the cooling load and, therefore, indirectly with people density, providing effective ventilation demand control at no added cost.

For other applications, a control system that modulates outdoor air intake to maintain a maximum space CO_2 concentration may be considered. CO_2 concentration is indicative of indoor air quality for spaces whose primary sources of indoor air pollution are the occupants themselves.

Ventilation for Multiple-Zone Systems

There is much confusion concerning how to properly control ventilation for central systems that serve multiple zones, each with varying, and possibly conflicting, air conditioning load and ventilation requirements. The system can only supply a given percentage of outdoor air to all spaces. But one space may require a high air-flow rate because it has a high cooling load, while another may receive only a small rate due to low cooling loads. At the same time, the ventilation requirements of the two spaces may be the opposite. That is, space with the high cooling load such as a perimeter office with large windows may require only a small amount of outdoor air. However, a lightly loaded space such as an interior conference room may require high outdoor air rates. The situation is further complicated for VAV systems where the supply air rates are constantly changing.

To practically address this problem the following approach is suggested for central, multiple-zone systems, particularly VAV systems.

Transfer air

For some spaces ASHRAE standards allow minimum ventilation supply air quantity to be met by transferring air from adjacent spaces that are overventilated. For instance, a conference room may be exhausted with make-up air transferred from an adjacent occupied space that is supplied with enough outside air to meet the requirements of both spaces.

The standards restrict this approach to spaces that have exhaust air fans. Logically, the concept could be extended to spaces that are supplied with air transferred from other spaces. This would, for instance, allow plenum air supplied by fan-powered VAV boxes to meet ventilation air quantities, provided other spaces were open to the return plenum.

Use the transfer air approach for zones that require high ventilation rates using transfer exhaust fans or fan-powered mixing boxes. Base peak air quantities on one-half the design occupancy if the space is typically occupied at peak for less than 3 hours.

Dilution approach

Assuming all zones are at design conditions, the dilution approach detailed in section 6.1.3.1 of ASHRAE Standard 62 can be used to determine the design peak quantity of outdoor air. The dilution equation from Standard 62 can be expressed as:

$$Vot = Von \div (1 + X - Z)$$
$$X = \text{Von} \div Vst$$
$$Z = Voc \div Vsc,$$

where

- Vot = the required overall outdoor air intake;
- Von = the sum of outdoor air intakes required for all zones served by the system;
- X = the uncorrected fraction of outdoor air;
- Z = the fraction of outdoor air for the "critical zone";
- Vst = the total supply volume of central fan (outdoor air plus recirculated air);
- Voc = the outdoor air supply required for the critical zone;
- Vsc = the total supply air volume to the critical zone;
- Critical Zone = the zone with the highest value of Z.

Critical zone

In a zoned HVAC system, the ventilation zone requiring the highest static pressure is called the critical zone. The supply fan delivers the pressure necessary to satisfy the ventilation required for the critical zone at that specific time, thus minimizing costly duct system over pressurization commonly found in conventional systems.

By satisfying the critical zone, all the other zones will also be satisfied. In fact, this control method inherently guarantees that no zone will be starved for air.

The "critical zone" is selected from those zones that are not ventilated with transfer fans. Conference rooms, for example, would not be used as the critical zone. The typical critical zone would be an interior office space that has a low cooling load but an occupant density similar to other zones. The dilution equation will then establish the minimum ventilation rate required for the building. For VAV systems, this same rate is used regardless of actual system supply volume since dynamic, real-time calculation of ventilation rate is impractical. CO_2 control systems are recommended to reduce this rate when the building is only partially occupied.

Note: If the transfer air concept is adopted, the dilution equation need not be used. The overall outdoor air rate required would simply be the sum of the outdoor air rates required for each space. This is because any excess air supplied to one space is transferred via the return system to other zones that may be under ventilated.

VAV minimum setpoint

Proper setpoints for minimum supply air volumes on VAV boxes are a controversial issue. The following are several design approaches that have been used successfully:

- No primary air minimum volume need be maintained if:
  1. boxes are fan-powered and the fan operates either continuously (series type), or
  2. when primary air volumes are reduced to below the ASHRAE standard ventilation rate (parallel type), or
  3. the minimum outdoor air rate for the whole system is maintained.

- For those who only partially agree with the transfer air concept, primary air minimum volume settings should be equal to the minimum outdoor airflow rate required for the zone. (This ignores the fact that the primary supply air may not be 100% outdoor air. Again, maintain the overall minimum for the system back at the main fan.)
- For those who do not believe in the use of transfer air at all, set minimum rates using the same concept used to develop the dilution equation:

$$Vzm = Voz \div Z,$$

where

- Vzm = the zone minimum VAV box setting;
- Voz = the zone minimum required outdoor air;
- Z = the critical zone ratio of outdoor air to overall supply air (as defined earlier).

This equation, which assumes that the dilution equation was used to establish the overall quantity of outdoor air, accounts for the dilution of return air to minimize volume settings.

The conservatism in using this approach can be seen by applying it to the critical zone. Minimum volumes are equal to maximum design volumes, which means the zone supply air would be constant volume. Some form of reheat would be required for temperature control.

Piping insulation

Many local codes now require all piping associated with HVAC systems be thermally insulated. The values in Table 10-1 show the minimum thickness of insulation having conductivity falling in the range listed, when tested at the mean rating temperature listed for each fluid design temperature range. The conductivity is typical of fiberglass and most elastomeric foam insulation.

Exceptions

Insulation is not regulated in the following cases:

- Piping that is factory installed within equipment. The intent here is to exempt piping within equipment whose energy

Table 10-1 Minimum pipe insulation (in.).

Fluid design operating temperature range °F	Insulation conductivity			Nominal pipe diameter (in.)				
	Conductivity rangeBtu-in/h-ft3°F	Mean rating tempeature °F	Run outs (a) up to 2	1 and less	1-¼ to 2	2-½ to 4	5 & 6	8 & up
Heating Systems (Steam, Steam Condensate, and Hot Water)								
Above 350	0.32–0.34	250	1.5	2.5	2.5	3.0	3.5	3.5
251–350	0.29–0.31	200	1.5	2.0	2.5	2.5	3.5	3.5
201–250	0.27–0.30	150	1.0	1.5	1.5	2.0	2.0	3.5
141–200	0.25–0.29	125	0.5	1.5	1.5	1.5	1.5	1.5
105–140	0.24–0.28	100	0.5	1.0	1.0	1.0	1.5	1.5
Domestic and Service Hot Water Systems (b)								
105 and Greater	0.24–0.28	100	0.5	1.0	1.0	1.5	1.5	1.5
Cooling Systems (Chilled Water, Brine, and Refrigerant) (c)								
40–55	0.23–0.27	75	0.5	0.5	0.75	1.0	1.0	1.0
Below 40	0.23–0.27	75	1.0	1.0	1.5	1.5	1.5	1.5

Notes:

(a) Runouts to individual terminal units not exceeding 12 ft in length.

(b) Applies to recirculating sections of service or domestic hot water systems and first 8 ft from storage tank for non-recirculating systems.

(c) The required minimum thickness does not consider water vapor transmission or condensation. Additional insulation, vapor retarders, or both, may be required to limit water vapor transmission and condensation.

performance is tested, and piping losses are ostensibly accounted for in the ratings.

- Piping conveying fluids that have design operating temperatures between 55 and 105°F, such as typical condenser water piping.
- Piping that conveys fluids that have not been heated or cooled using fossil fuels or electricity. This exception is intended to cover gas piping, cold domestic water piping, waste and vent piping, rain water piping, etc. The exception is due to these pipes carrying fluids with operating temperatures outside the 55 to 105°F range but for which no energy was consumed to bring the fluids to these temperatures. While insulating such piping will have no energy impact and is not required, it may be desirable in some cases, and possibly required by building codes, to prevent condensation.
- Where it can be shown that heat gain or heat loss to or from the piping will not increase building energy costs. Examples of piping falling into this exception category are condensation drains, liquid and hot gas refrigerant lines on AC units, and liquid lines on heat pumps.

Duct insulation

As seen in Table 10-2, ductwork and air plenums also require thermal insulation. Caution should be exercised if "cold-air" systems are used. They may supply air 10 to 15°F colder than

Table 10-2 Minimum Duct insulation

Duct location	Temperature difference (°F)	Insulation R-value
Exterior of building	All	8.0
Inside of building envelope or in unconditioned spaces		None required
	DTe £ 15 —	40
	>DTe > 15 —	3.3
	DTe > 40 —	5.0 t

the usual supply air temperature. Even in the usual temperature range, condensation can occur if insulation is unduly compressed or if there are openings or gaps in the vapor regarding envelope.

Requirements are divided into two primary categories:

- Ducts that are outdoors, exposed to weather.
- Ducts that are either in unconditioned spaces (enclosed but not within the building envelope) or in conditioned spaces (within the building envelope). This second category includes semienclosed spaces such as vented attics and crawl spaces.

Insulation requirements for enclosed ducts depend on a value of ΔT, defined as the temperature difference between the air conveyed in the duct and the ambient air surrounding the duct, both at design conditions. In many cases, determining these temperatures will require some judgment on the part of the designer, particularly for ambient temperatures of unconditioned spaces. A rough estimate of supply air, return air, and unconditioned space temperatures will usually suffice since each requirement is for a broad range of temperature differences.

The R-values listed in Table 10-2 are for the insulation material as installed, excluding air film resistance, but including the effect of the compression of duct wraps as they are typically installed. Duct wraps are generally assumed to be compressed to 75% of their nominal thickness when they are installed. Material conductivity is as measured in accordance with ASTM C518-85 at 75°F mean temperature. Common materials that meet the R-value levels in Table 10-2 are shown in Table 10-3.

Default duct insulation values

For supply and return ducts inside of the building envelope or in unconditioned spaces, application of Table 10-2 is based on the determination of a ΔT. Design air temperatures for supply ducts will vary from one system to another, but typically are in the range of 55°F to 60°F for cooling. Using these temperatures, default values of ΔT for a variety of supply and return duct locations were calculated and are presented in Table 10-4. In Table 10-4 the default ΔT (and consequently the de-

Table 10-3 R-values for common duct insulation materials.

R-value	(hr-°F-ft2)/Btu nominal thickness	Typical materials
3.31–1/2 in.	1/2 to 1-1/2 lb/ft3	fiberglass duct wrap
	1 in. 3/4 to 3 lb/ft3	fiberglass duct liner
	1-1/2 in. 1/2 lb/ft3	fiberglass duct liner
1 in.		fibrous glass ductboard
1 in.		insulated flexible duct
5.03 in.	1/2 lb/ft3	fiberglass duct wrap
2 in.	3/4 to 1-1/2 lb/ft3	fiberglass duct wrap
	1-1/2 in. 3/4 to 2 lb/ft3	fiberglass duct liner
	1 in. 3 lb/ft3	fiberglass duct liner
6.53 in.	1/2 to 1-1/2 lb/ft3	fiberglass duct wrap
2 in.	3/4 to 1-1/2 lb/ft3	fiberglass duct liner
	1-1/2 in. 2 to3 lb/ft3	fiberglass duct liner
8.04 in.	1/2 to 3/4 lb/ft3	fiberglass duct wrap
3 in.	1 to 1-1/2 lb/ft3	fiberglass duct wrap
3 in.	3/4 to 1 lb/ft3	fiberglass duct liner
2 in.	1-1/2 to 3 lb/ft3	fiberglass duct liner

From ASHRAE Standard 90.1- 1989 User's Manual.

fault minimum required insulation R-value) are determined by the design outdoor-air dry-bulb temperature. For buried ductwork, use the design ground temperature to determine the insulation requirements.

Air system balancing

Air systems must be balanced by first adjusting fan speed to meet design flow. Damper throttling alone may be used for balancing fans only under two conditions:

- with motors of 1 hp or smaller,
- where throttling will increase fan power by no more than ⅓ hp above that required if the fan speed were adjusted.

Hydronic system balancing

Hydronic systems must be balanced by adjusting pump speed or by trimming the pump impeller to meet design flow requirements. Valve throttling alone may be used for final balancing only in the following cases:

Table 10-4 Default Ts and R-values for supply and return ducts in the building envelope or in unconditioned spaces. (a)

			Cooling (b)	
Duct type	**Location**	**$\Delta T \geq 15$ none required**	**$40 > \Delta T >$ 15r-3.3**	**$\Delta T > 40$R-5.0**
Supply	Ventilated attic		<85	>85
	Unvented attic		<80	>80
	Shaft or crawlspace		<95	>95
	Plenum		all	
	Buried (see note c)	<70	>70	
Return	Ventilated attic	<82	82-107	107
	Unvented attic	<77	77-102	102
	Shaft or crawlspace	<92	92-117	117
	Plenum wall	<92	92-117	117
	Buried	all		

Notes:
(a) Assumptions are as follows: 55°F/77°F cooling supply/return temperatures; ventilated attics run 10°F warmer than outside air temperatures during peak cooling; unvented attics run 15°F warmer than outside air temperatures during peak cooling; shafts and crawl space temperatures are equal to outside air temperatures during peak cooling.
(b) Entries in the table are the design cooling dry-bulb temperatures (°F). The 2.5% values from the ASHRAE Fundamentals Handbook should be used.

- Pumps with motors of 10 hp or less, if valve throttling results in an increase in pump power of no more than 3 hp above that required if the impeller were trimmed.
- To reserve pump capacity to overcome future fouling in open circuit piping systems (cooling tower systems).

Throttling losses may not exceed the pressure drop expected for future fouling.
- Where it can be shown that throttling will not increase overall energy costs.

This last point may be the case for some condenser water systems since an increase in flow rate can reduce chiller energy usage enough to offset increased pump and cooling tower energy usage. A detailed analysis, including both part-load and full-load operating conditions, must be made.

Fan power can be calculated from the following equation:

$$W = 746 \times \left(\frac{\text{BHP}}{\text{Nm} \times \text{Nd}}\right)$$

where

- *W* = the fan power in watts;
- *BHP* = the fan brake horsepower, which is the power required at the fan shaft, measured in horsepower;
- *Nm* = the fan motor efficiency
- *Nd* = the drive efficiency, such as belt drives and, where applicable, variable speed drives.

The fan brake horsepower requirement can be found in manufacturer's catalog data or generated by their proprietary computer programs. Where data are not available, assume that fan brake horsepower is equal to fan motor horsepower.

Motor efficiencies will vary by motor type, speed, and size. Table 10-5 provides typical motor efficiencies that should be used if actual data are not known. Seasonal equipment efficiency will also have an effect; Table 10-6 provides some heating and cooling equipment data.

Belt drive efficiencies vary depending on the number and type of belts. If better data are not available, assume belt drives are 97% efficient. Variable speed drive efficiencies, which vary with drive type, can be obtained from the manufacturer.

Estimating heating requirements

In accordance with natural law, heat will always flow in a direction from hotter to cooler. For this reason, any structure

Table 10-5 Typical motor efficiencies.

Nameplate rating (hp)	Standard motor	High efficiency motor
1/20	35	—
1/10	35	—
1/8	35	—
1/6	35	—
1/4	54	—
1/3	56	—
1/2	60	—
3/4	72	—
1	75	82.5
1-1/2	77	84.0
2	79	84.0
3	81	86.5
5	82	87.5
7-1/2	84	88.5
10	85	89.5
15	86	91.0
20	87	91.0
25	88	91.7
30	89	92.4
40	89	93.0
50	89	93.0
60	89	93.6
75	90	94.1
100	90	94.1
125	90	94.5
150 and up	91	95.0

Open Motors, 1800 RPM synchronous speeds, nominal efficiencies.

with an inside temperature that is higher than the ambient outside temperature will lose heat. An entire branch of science known as thermodynamics studies the pattern and relationship of heat flow to temperature differences. According to thermodynamic principles, then, a structure with an inside temperature of 70 degrees Fahrenheit, on a day when the outside air temperature is less than 70 degrees (say 35 degrees), will experience heat loss due to the inside air seeking pathways through which to escape to the cooler outside air.

Table 10-6 Heating and cooling equipment data.

	Low	Medium	High	Very high
HEEF				
Gas furnace	0.50	0.65	0.80	0.90
Oil furnace	0.50	0.65	0.80	0.90
Heat pump (HSCOP)	1.6	1.9	2.2	2.5
Electric furnace	1.0	1.0	1.0	1.0
Electric baseboard	1.0	1.0	1.0	1.0
CEEF				
Heat pump (SEER)	7.25	8.75	10.25	11.75
Air conditioner (SEER)	6.0	8.0	10.0	12.0

The manner in which the relative warmth or coolness of a solid, liquid, or gas is described is through the use of one of two scales—degrees Fahrenheit or degrees centigrade. Both of these scales are based on the boiling and freezing temperature of water. In degrees Fahrenheit, water freezes at 32 degrees and boils at 212 degrees. According to the centigrade scale, water freezes at 0 degrees and boils at 100 degrees.

When dealing with heating systems it is likely that both of these temperature scales will be encountered. To convert from one scale to the other follow these procedures:

- To convert from Fahrenheit to centigrade, subtract 32 from the Fahrenheit temperature and divide the result by 1.8.
- To convert from centigrade to Fahrenheit, do the reverse: multiply the centigrade temperature by 1.8 and add 32.

Sizing a heating system

Since a structure, no matter how well-insulated, is bound to lose a certain amount of heat, the heating system for the structure must be capable of replacing that heat loss. Heat loss occurs through components of the building that have direct contact with the outside, such as walls, windows, doors, floors, and ceilings. The amount of heat lost through any of these components will be affected by the size of the component, its composition, and the degree to which each component is insulated.

- *British Thermal Units (Btu).* A Btu is a unit of measure equal to a quantity of energy and is normally associated with the transference or production of heat. The most commonly used definition of a British thermal unit is the amount of heat required to raise the temperature of 1 pound of water 1 degree. Manufacturers of heating systems rate the output of their systems in terms of Btuh (British thermal units per hour). By expressing the anticipated or actual heat loss of a building in the same terms, selecting a properly sized heating plant becomes much easier.
- *Degree Days.* One of the methods the heating industry uses to determine the severity of the heating seasons in a given area of the country is to calculate the number of degree days during the season. A degree day is a unit of measurement used in the heating and cooling industry in conjunction with reference temperatures to determine the approximate number of days a heating or cooling plant will need to run during a season.

A degree day is, essentially, 1 degree Fahrenheit difference between the actual average outside air temperature and a reference temperature representing a temperature at which people would experience minimum comfort. For example, the fuel industry normally uses 65 degrees Fahrenheit as the reference temperature for calculating fuel consumption during winter months. If the actual average temperature was 37 degrees, then $65 - 37 = 28$ degree days.

Determining heat loss

A popular method for determining heat loss involves the use of geographic factors that have been calculated to correspond to specific areas of the United States. See Table 10-7.

Table 10-8 presents numbers to be used as multipliers. These multipliers correspond to both the geographic factors and to the building's general construction.

Calculating the amount of a structures heat loss in terms of Btu is accomplished with the following steps:

Table 10-7 Approximate geographic adjustment factors for the United States.

Location	Heating factor in degrees Fahrenheit
Alabama	10
Anchorage, Alaska	–20
Fairbanks, Alaska	–50
Flagstaff, Arizona	–10
Phoenix, Arizona	25
Arkansas	5
California	30
Colorado	–15
Connecticut	0
Delaware	0
District of Columbia	0
Florida	25
Georgia	15
Hawaii	62
Idaho	–10
Illinois	–10
Iowa	–15
Kansas	–10
Kentucky	0
Louisiana	20
Maine	–15
Maryland	0
Massachusetts	–5
Michigan	–10
Minnesota	–25
Mississippi	10
Missouri	–5
Montana	–28
Nebraska	–10
Nevada	–5
New Hampshire	–15
New Jersey	0
New Mexico	0
New York	–5
North Carolina	05
North Dakota	–30
Ohio	–5
Oklahoma	0
Oregon	10

Table 10-7 Approximate geographic adjustment factors for the United States (continued).

Location	Heating factor in degrees Fahrenheit
Pennsylvania	0
Puerto Rico	68
Rhode Island	0
South Carolina	15
South Dakota	–20
Tennessee	0
Texas	10
Utah	–10
Vermont	–10
Virginia	15
Washington	0
West Virginia	0
Wisconsin	–15
Wyoming	–15

Table 10-8 Heat loss multipliers

Construction component	–30	–20	–10	0	10	20	30
Walls—without insulation	26	24	21	18	15	13	10
Walls—with insulation	13	12	11	9	8	6	5
Windows and doors with insulation	170	150	130	120	105	85	65
Window and doors without insulation	80	70	60	55	60	40	30
Ceilings with uninsulated attic above	32	29	25	22	19	15	12
Ceilings with insulated attic above	10	9	8	7	6	5	4
Floors over concrete slabs	10	9	8	7	6	5	4
Floors over heated basements	0	0	0	0	0	0	0
Floors over insulated crawlspaces	5	5	4	4	3	3	2
Floors over un-insulated crawlspaces	14	13	11	10	9	7	6

1. Calculate the perimeter of the structure (the length of the exposed walls). For example, assuming a structure with four walls each measuring 50 feet, the perimeter would be 200 feet.
2. Multiply the perimeter figure by the sum of the heights of the ceilings on each level with the exclusion of the basement level. For example, if a building has two stories, each with 8-foot ceilings, the multiplier would equal $8 \times 2 = 16$. And $200 \times 16 = 3200$ square feet.
3. Find the geographic factor in column 3 of Table 10-7 that corresponds with the location of the building. For example, the factor for Illinois is –10.
4. Referring to Table 10-8, find the column headed by the geographic factor (–10). For each component of the structure, locate the appropriate multiplier within that column. For example, assume the exposed walls are insulated; thus, the multiplier will be 11.
5. Multiplying the total square footage from Step #2 by the multiplier located in Step #4 will yield the total heat loss in terms of Btu's. So $3200 \times 11 = 35{,}200$.
6. Calculate the area of all windows, separating single-pane windows from insulated windows.
7. Calculate the area of all doors, again separating insulated doors from non-insulated doors.
8. Calculate the area of the ceiling on the top level of the building.
9. Calculate the area of all flooring on the ground level of the structure, including floors over basements and crawlspaces.
10. Referring again to Table 10-8, locate the appropriate multiplier for each of these components and perform the calculations to convert to Btu's.
11. Adding the total heat loss expressed as Btu's for each of the components listed here will provide the total heat loss (in Btu's) for the entire structure.

Once the total heat loss for all areas of the structure is converted to Btu's, sizing a heating plant to meet the needs of the building is only a matter of matching the heat loss (in Btu's) to the heat plant's Btu output.

Estimating cooling system requirements

An adequate cooling system (for any building) actually has two functions. It should have the capacity to cool the structure to a minimum of 75 degrees Fahrenheit indoor temperature while at the same time maintaining a relative humidity of no more than 50 percent inside the building. Although some people find higher temperatures and humidity tolerable, studies have confirmed 75 degrees Fahrenheit at 50-percent relative humidity to be the upper end of most people's comfort level.

Sizing a cooling system

In determining the appropriate cooling system for a building, there are a number of factors to be considered. Some of these factors are the same as for sizing a heating system, such as exposed wall area and top floor ceiling area. However, calculating cooling requirements also takes into consideration some factors yet to be discussed, such as the number of people who occupy the structure.

Because humans maintain a constant average body temperature of 98.6 degrees Fahrenheit, radiated heat from the occupants of a building will increase the inside temperature. Likewise, the human body's major component is water, and with every breath taken, moisture is released into the atmosphere, thus increasing the relative humidity of the air inside the structure. In residential construction, the home's kitchen is a daily source of additional heat and humidity. For a restaurant, the heat and humidity contributed by the kitchen will be an even more significant factor to consider.

A list of geographical factors exists for cooling calculations, just as it does for heating. For cooling however, there are two factors which are given; the geographic temperature factor and the geographic humidity factor. These factors are statistical averages based on accumulated weather data and are presented in Table 10-9 below. The humidity factor is actually the ideal temperature (75 degrees Fahrenheit) which has been adjusted to account for the location's relative humidity during summer months. Any location with a humidity

Table 10-9 Temperature and humidity factors used in calculating cooling requirements.

Location	Temperature factor	Humidity factor
Alabama	95	78
Anchorage, Alaska	67	59
Fairbanks, Alaska	75	73
Flagstaff, Arizona	90	65
Phoenix, Arizona	105	76
Arkansas	95	78
California	98	71
Colorado	95	65
Connecticut	95	75
Delaware	95	78
District of Columbia	95	78
Florida	95	79
Georgia	95	78
Hawaii	84	73
Idaho	95	65
Illinois	96	76
Iowa	95	78
Kansas	100	78
Kentucky	95	78
Louisiana	95	80
Maine	90	73
Maryland	95	78
Massachusetts	93	75
Michigan	95	75
Minnesota	93	73
Mississippi	95	78
Missouri	100	76
Montana	95	76
Nebraska	95	78
Nevada	95	65
New Hampshire	90	73
New Jersey	95	75
New Mexico	95	70
New York	93	75
North Carolina	95	78
North Dakota	95	73
Ohio	95	76
Oklahoma	101	77

Table 10-9 Temperature and humidity factors used in calculating cooling requirements (continued).

Location	Temperature factor	Humidity factor
Oregon	90	68
Pennsylvania	95	76
Puerto Rico	87	79
Rhode Island	93	75
South Carolina	95	78
South Dakota	95	75
Tennessee	95	78
Texas	100	78
Utah	95	65
Vermont	90	73
Virginia	95	78
Washington	89	65
West Virginia	0	
Wisconsin	95	75
Wyoming	95	65

factor of 75 or less requires no additional consideration due to humidity conditions.

Windows and doors with glass panels are building components that play a critical role in determining the cooling load of the structure. The directional alignment of each window, combined with the window treatments (blinds, curtains, sheers, etc.), used on them, will have an effect on how much radiant heat is added to the structures cooling load. Table 10-10 provides factors corresponding to window alignment, by which the total glass area should be multiplied.

Another area of the structure that will impact the cooling load is the uppermost ceiling, although the exact degree depends on the type of roof, the color of the roofing material, whether or not there is attic space under the roof, whether or not that space is properly ventilated, and how much insulation is in the attic space. Table 10-11 presents the factors associated with these variables.

The following procedures will allow the contractor to determine required cooling needs for a given structure:

Table 10-10 Glass area multipliers corresponding to location.

Directional alignment of glass	Factor
West or East with no shading	117
West or East with shading	80
South with no shading	65
South with shading	45
North (or completely shaded)	40

Shading refers to window treatments on the inside (blinds, heavy drapes, etc.), or objects outside (trees, bushes landscape walls etc.) which partially or completely block direct sunlight from passing from the window to the interior of the structure.

Doors with no glass panels are assigned a factor of 20.

1. Multiply the estimated average number of occupants for the building by 400 to derive the number of British thermal units generated per hour by the occupants. For example, if a building is designed as a single family dwelling it may be estimated to normally house four people.

$$4 \times 400 = 1600 \text{ Btuh}$$

Table 10-11 Cooling load factors corresponding to upper most ceiling area.

Area above upper most ceiling	Insulation of space above ceiling	With proper ventilation	Dark color roof surface	Light color roof surface
Attic	No insulation	9	13	10
	Adequate insulation	2	3	3
No attic (BUR*)	No insulation	-	23	18
	Adequate insulation	-	8	7

* BUR = Built Up Roofing

2. The total perimeter of the structure multiplied by the total height of the ceilings equals the total area of the exposed walls. The total square footage of average insulated exposed walls is multiplied by a factor of 2 to determine this portion of the cooling load (use a factor of 4 for inadequately insulated walls or walls with no insulation). Using the previous example of a 2 story with 4 walls, each measuring 50 feet and 16 feet of ceiling height the calculation would be:

$$(4 \times 50 = 200)\ 200 \times 16 = 3200 \text{ square feet}$$

$$3200 \times 2 = 6400 \text{ Btuh}$$

3.
   a. Categorize the number of windows and doors which have glass panels according to the direction in which they are aligned.
   b. Calculate the total glass area facing each direction.
   c. Referring to Table 10-10, multiply the total glass area allocated to each direction by the corresponding factor.

   In the example above, if the house had total glass area of 325 square feet with 180 square feet of that total facing North and the other 145 square feet facing South, with no shading, the calculations would be:

$$(180 \times 40) + (145 \times 65) = 16{,}625 \text{ Btuh}$$

4.
   a. Calculate the area of the upper most ceiling in the building.
   b. Referring to Table 10-11, attain the factor which corresponds to the conditions which pertain to the upper most ceiling. For this example, assume an attic space with proper ventilation and insulation with a light colored roof, which equates to a factor of 2.
   c. Multiply the area of the ceiling ($50 \times 50 = 2500$ square feet) by 2 to obtain the Btuh load contributed by the ceiling.

$$2500 \times 2 = 5000 \text{ Btuh}$$

5. To account for the portion of the cooling load attributable to the kitchen, an estimator of 1500 Btuh is used.

6. Add the Btuh figures from each of the steps above to arrive at the total Btuh cooling load. In this example: 1600 + 6400 + 16,625 + 5000 + 1500 = 31,125
7. Locate the appropriate geographic temperature factor in Table 10-9. If this factor is greater than 95 or if an inside temperature less than 75 degrees Fahrenheit is desirable, a correction to the factor must be made. Subtract the 75 (or the lower, desired temperature) from the given geographic factor. Assuming a given factor of 100, then 100 − 75 = 25. Next, subtract 20 from the result (25) to derive the additional number of degrees of cooling required (25 − 20 = 5). For every 5 degrees of additional cooling load, add 25 percent to the total Btuh figure arrived at in step No. 6 (31,125 Btuh).

$$31{,}125 \times 0.25 = 7781$$

$$31{,}125 + 7781 = 38{,}906 \text{ Btuh}$$

8. A similar correction is necessary for areas of high relative humidity. Fortunately, there are only few locations with a geographic humidity factor greater than 75. For factors from 76 to 78, increase the total Btuh figure from Step #6 (31,125) by 5 percent. If the factor is above 78, increase the figure from Step #6 by 10 percent.

Once the total Btuh figure has been determined in Step No. 6, and any necessary corrections are made, the corrected total Btuh figure (38,906 Btuh for example) would be the approximate size of the air conditioning unit appropriate for the building in this instance.

Solving ventilation problems

A poorly ventilated attic space is a source of problems throughout the entire year. Even if the space has adequate insulation, there will be some degree of air infiltration summer or winter. During the cold months, warm air rising from the living area will be laden with moisture from bath and kitchen uses as well as the moisture exhaled by the occupants. Without proper ventilation, the moisture in this warm air will make contact with the cold underside of the roof, condense, and cause ruined insulation, rotted wood, damaged roofing materials, and blistered inside ceilings.

During hot weather months, the build up of heat in an unvented attic space will decrease the life of asphalt roofing and impair the efficiency of air conditioning units. With trapped hot air in the attic reaching temperatures of 135 degrees Fahrenheit or more, a drywall ceiling will allow a portion of this heat to penetrate into the living space. To calculate the heat gain into the living area, begin by subtracting the desired indoor air temperature from the actual temperature of the attic. For an attic at 135 degrees and a desired indoor temperature of 75 degrees, this is a difference of 60 degrees. For a building with 3200 square feet of living area, multiply 3200 by 0.12 Btu times the number of degrees difference from the inside temperature to the temperature of the attic. The calculation would appear as:

$$3200 \times 0.12 \times 60 = 23{,}040$$

That is, without proper ventilation, a structure with the conditions described above would be adding 23,040 Btuh per square foot of living space (almost two tons of cooling requirement).

Passive ventilation

Any means of ventilating an area without the use of a motor driven device falls into the category of passive ventilation. Most passive ventilation takes the form of different-sized and -shaped vents installed at specific locations such as gable ends, soffits, or ridge lines. The important concept to grasp with ventilation is that of intake and exhaust. Just like a person breathing, a structure must breath also; there must be a flow of air coming into the attic from one vent site that will flow to an exhaust vent at another location. This flow of air will cause an exchange of air, with the constant movement removing heat and moisture.

- *Soffit and ridge vents.* Normally, a series of rectangular soffit (the area under the eaves) vents (Fig. 10-3) along two sides of the structure work in conjunction with a ridge vent, which as its name implies runs along the ridge of the roof. Air entering through the soffit vents flows through the space exiting from the ridge vents. Soffit vents are also used in conjunction with gable end vents.
- *Gable end vents.* Generally rectangular or triangular in shape, gable end vents placed opposite each other set up a

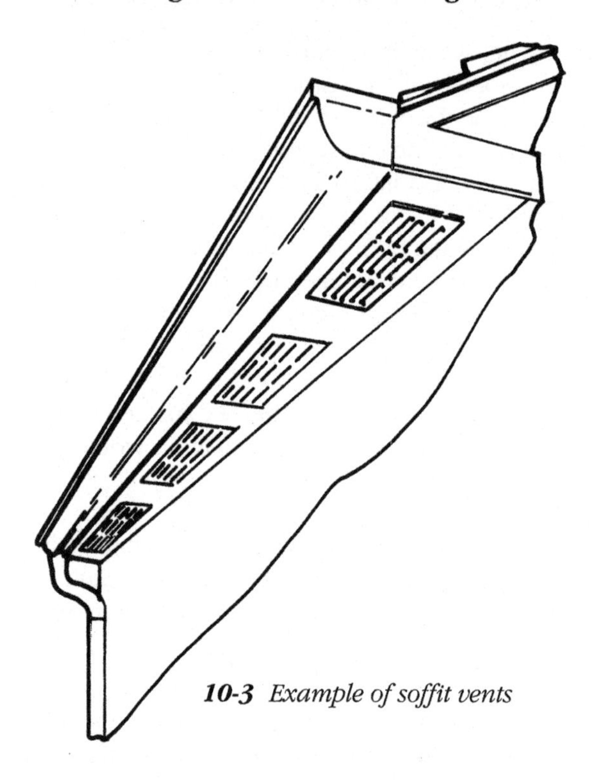

10-3 *Example of soffit vents*

stream of cross ventilation. The intake of air from soffit vents is pulled up and through the space by the cross ventilation stream of the gable end vents.

- *Turbine.* A wind-driven turbine (Fig. 10-4) placed slightly down from the ridge of the roof will work well in conjunction with soffit vents. As the wind blows, the turbine rotates pulling air up the flu of the turbine. This creates a flow of air fed from the soffit vents.

Power ventilation systems

These systems combine the passive intake vents described above with a motor driven exhaust unit to create a positive exchange. Power ventilation systems can be manually controlled or automatic. Equipped with a thermostat and a humidistat to control the units function, the exhaust unit will automatically engage when the temperature or humidity is higher than preset acceptable levels. Once these two factors have been re-

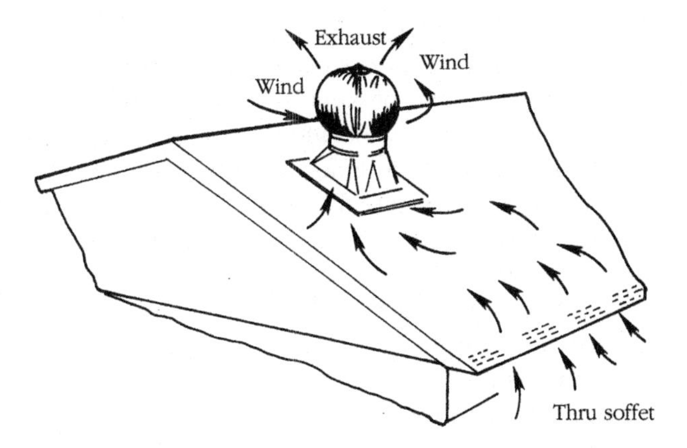

10-4 *A wind-driven turbine used for passive ventilation*

duced to acceptable limits, the unit will shut itself down. Because power units are not subject to failing winds, they will be more efficient in exchanging the attic air.

- *Turbine.* Appearing much the same as the passive turbine pictured in Fig. 10-4, the power turbine functions passively (with the wind) and actively with the help of a motor to turn the turbine blades.
- *Cupola.* Often when a power vent (or fan) is to be used on the roof, the construction of a cupola astride the ridge is not only attractive but practical for housing and protecting the fan as well as sealing and protecting the opening in the roof.

In many cases, the potential of the power ventilator is great enough to pull air from the first floor of a two-story home up through the attic and out. On moderately hot days, this type of full building ventilation can be enough to cool the structure without use of the air conditioner or at the very least will assist the air conditioner by lessening the cooling load.

Calculating required ventilation

The area to be vented (the attic) should be measured to determine its square footage. This holds true whether passive or power ventilation is to be used. With passive systems, the square foot figure for the attic is then divided by 150 if there is no vapor barrier facing the living space, or 300 if there is a vapor barrier in place. This calculation will yield the required number of square feet of ventilation. Half the number of square feet dedicated to ventilation must be designed for intake, with the other half designed for exhaust.

The fans of power systems are sized by numerical designations according to the volume of air the fan is capable of moving in one minute's time, referred to as a cubic-foot-per-minute (CFM) rating. These CFM rating numbers allow easy comparison and selection of units. A quick, accurate way to determine the appropriate cubic foot per minute requirement of an attic space is as follows:

1. Multiply the square footage of the attic by 0.70 to obtain the CFM minimums for the area.
2. For dark colored roofs increase the CFM minimums by 15 percent.

The importance of good attic ventilation beneath the roof cannot be overemphasized. Such movement of air will prevent or inhibit the condensation of moisture on the undersurface of the shingles or shakes, or on the roof deck and rafters. Vents should be provided at the soffits (eaves) as well as at gable ends (screened to prevent ingress of insects) or preferably the ridge lines with cross-ventilation desirable. A rule of thumb for adequate ventilation, as with asphalt shingles, is that the ratio of total net free ventilation area to the area of the attic should not be less than 1 : 150, with compensation made for screens or louvers over vent apertures. Table 10-12 provides adjusted vent sizes which compensate for screening materials.

If ice-damming is a potential problem, or if reverse condensation is likely to occur, the cold weather roof system should be used in conjunction with horizontal strapping. Remember that ventilation must be provided at the eaves and at the peak. The principle of the cold weather roof system is to allow a constant flow of cold air above the insulation but below the roofing material. With other roofing systems, ice buildup along the eaves can be a problem. Heat escapes from the insulation and melts snow, which runs down the roof to the cold overhangs where it again freezes, backing up under

Table 10-12 Vent size adjustments due to screening.

Screen or covering	Opening size
¼ inch hardware cloth	1 times the net vent area
¼ inch hardware cloth with rain louvers	2 times the net vent area
8-mesh screening	1¼ times the net vent area
8-mesh screening with rain louvers	2¼ times the net vent area
16-mesh screening	2 times the net vent area
16-mesh screening with rain louvers	3 times the net vent area

roofing material, where it frequently will melt again and possibly penetrate the roof system. A properly installed, vented cold weather roof eliminates this problem. Venting space should be sufficient to allow a free flow of air from eave to roof top.

Ventilation for crawl spaces

The minimum ventilation for an unheated crawl space should be provided by two vents opposite each other. Remember, at least two vents must be present in order to provide intake and exhaust avenues for the flow of air. The amount of ventilation needed is again determined by dividing the total square footage of the crawl space. If the crawl space has no moisture seal, the ventilation requirements are calculated by dividing the square footage of the area by 150. Depending on the size of the crawl space, two vents may become impractical due to the size each vent would need to be in order to fulfill the 1 : 150 ratio required for proper ventilation. For this reason, it is usually advisable to distribute the ventilation requirements among four vents (one on each exposed side of the crawl space).

If the crawl space is fitted with an appropriate moisture seal (4 mil polyethylene sheeting min.), then the ratio changes to 1 : 1500 with the understanding that there must always be at least two vents, regardless of ratio calculations, in order to provide intake and exhaust channels.

Review questions

1. A VAV fan system includes a 60 hp supply fan and a 15 hp return fan. If local codes have a 50% kW at 50% cfm requirement, does this system have to meet it?

2. A 10-ton rooftop air conditioner has a single compressor with no unloading capability. Does this unit have to meet IPLV requirements?
3. A VAV system supplies cooling to exterior zones and interior zones in a building that has no ceilings. The supply ducts run exposed through the interior spaces. Do ducts require insulation?
4. Design load calculations on a zone of a reference school building have indicated a peak heating requirement of 360,000 Btu/h and a peak cooling requirement of 36 tons. What coil capacities should be used in the reference model?
5. A simulation tool has sized the airflow for a zone at 1.1 cfm/ft^2 and used that value in calculating the DECOS. The designer plans to use 20% more airflow as a safety factor and to handle the pick-up load. Is the calculated DECOS still acceptable?

11

Plumbing math

When laying out the plumbing requirements of a building, math becomes an indispensable tool. Planning the pipe runs throughout the structure and calculating water-supply needs, drainage systems, and ventilation system requirements all rely on mathematics to supply answers that will make the job easier and more cost-efficient.

Water supply systems

Every building needs a source of drinkable water that is sufficient to supply every fixture, inside and outside the structure, with either hot or cold water at the specific operating pressure required.

Public water supplies are the source of this drinkable water in most densely populated areas of the country today. Water that has been treated at these water works to meet the public health standards flows throughout the designated area of service in mains. Individual users can be connected to the public water system by tapping into these mains at the building site.

In many less densely populated areas, where a public water system is not available, wells drilled on the user's property and maintained by the property owner are an alternate source of water.

Water requirements

The use and supply of water is measured in terms of the water's rate of flow, which will normally be referred to in units of either gallons per hour or gallons per minute of flow. Regardless of the source of water supply (public or well), the approximate peak usage of the structure must be determined. A formula for determining water demand for a building involves adding up the number of fixtures in the structure and states that the total gallons per hour required equals the number of fixtures times 60:

$$\text{gph} = \text{fixtures} \times 60$$

For example, find the number of fixtures, add up the number of bathtubs and/or showers, toilets, sinks, garden faucets, utility tubs, clothes washers, and dish washers located on the site. As an example, an average residential structure with 12 fixtures would not be uncommon. Plugging this figure into the equation results in an approximate peak water demand of 720 gallons per hour.

$$\text{gph} = 12 \times 60$$

Flow measurement

Typically, engineering students take a course in fluid mechanics during which they learn one of the most fundamental equations of fluid mechanics—the Bernoulli equation. The Bernoulli equation is valid for simulating internal and external flows. Internal flow is flow inside a pipe or duct; external flow includes rivers, streams, or even raindrops falling from the sky.

One characteristic of the Bernoulli equation, as applied to internal flow, is that it is only valid when the situation is steady state. In other words, the fluid is incompressible and inviscid (i.e., no friction between the fluid and the object or pipe wall), and flow is along a streamline. In the real world, no flow situation perfectly matches these requirements. The Bernoulli equation, although serviceable only under ideal conditions, is taught because it is the precursor to the Energy equation.

The energy equation

The Energy equation is the Bernoulli equation with one additional term—head loss. Head loss (also known as energy loss) incorporates the effects of friction for internal flows. Since frictional effects are accounted for, the energy equation is able to simulate most actual, internal, steady flows where the fluid is relatively incompressible. Commonly used to predict pressure loss in a long pipeline, the Energy equation is:

$$Z_1 + P_1/\rho g + V_1^2/2g = Z_2 + P_2/\rho g + V_2^2/2g + h_L.$$

In this equation

- the Greek symbol ρ (rho) is used to represent density;
- Z_1 = upstream elevation;
- Z_2 = downstream elevation;
- P_1 = upstream pressure;
- P_2 = downstream pressure;
- V_1 = upstream velocity;
- V_2 = downstream velocity;
- h_L = energy (head) loss;
- g = acceleration due to gravity (32.174 ft/s^2) or (9.806 m/s^2).

This steady state energy equation applies to a fluid moving in a closed conduit and is based on the distance L (length) between two locations. The loss term h_L accounts for all minor (valves, elbows, etc.) and major (pipe friction) losses. Frequently, the flow rate and pipe diameter (or duct area) is known, so the equation is used to determine the downstream pressure at the end of a certain length of pipe. Since all of the losses (minor and major) are solved for, the Energy equation solves for the unknown. Major losses can be computed using either the Darcy–Weisbach method or Hazen–Williams method.

Darcy–Weisbach friction loss method

Major loss (h_f) is the energy (or head) loss (due to friction between the moving fluid and the duct) expressed in units of length. The Darcy–Weisbach friction loss equation is as follows:

$$h_f = f\,L/D\ V^2/2g, \text{ with } V = Q/A$$

For a non-circular duct, D is computed from the equation $D = 4A/P$. Within the equation

- h_f = major loss (ft);
- f = Moody Friction Factor;
- L = duct length (ft);
- D = duct diameter (ft);
- V = velocity (ft/s);
- g = acceleration due to gravity = 32.174 ft/s^2 = 9.806 m/s^2;
- Q = discharge (ft^3/s);
- A = duct area (ft^2);
- P = duct perimeter (ft).

In general, the Darcy–Weisbach method is considered more accurate than the Hazen–Williams method. Additionally, the Darcy–Weisbach method is valid for any liquid or gas; Hazen–Williams is only valid for water at ordinary temperatures (40–75° F).

Hazen-Williams friction loss method

Hazen–Williams is simpler than Darcy–Weisbach for calculations solving for flow rate, velocity, or diameter. This method is very popular, especially among civil engineers, since its friction coefficient (C) is not a function of velocity or duct diameter. Since the Hazen–Williams method is only valid for water flowing at ordinary temperatures (about 40–75°F) other liquids or gases should employ the Darcy–Weisbach method.

As with the Darcy–Weisbach method, major loss (h_f) is the energy (or head) loss due to friction between the moving fluid and the inside surfaces of the duct expressed in units of length. Major loss is frequently referred to as friction loss.
The equation is as follows:

$$V = kCR_hS, \text{ with } S = h_f/L,\ Q = AV, \text{ and } R_h = D/4 \text{ (for circular pipes).}$$

In this equation:

- V = velocity (ft/s);
- k = a unit conversion factor: k = 1.318 for English units (feet and seconds) or k = 0.85 for SI units (meters and seconds);
- C = Hazen–Williams coefficient;
- R_h = hydraulic radius = $D/4$ for circular pipe;
- S = energy slope (ft/ft);
- L = pipe length (ft);

- Q = discharge (ft^3/s);
- A = area (ft^2);
- D = pipe diameter (ft).

In order to simplify (and hopefully clarify) these equations, the following tables (Tables 11-1, 11-2, and 11-3) present some very useful information.

Kinds of flow

Flow is classified into open-channel flow and closed-conduit flow. Open-channel flow conditions occur whenever the flow-

Table 11-1 Minor loss coefficients (K has no units).

Fitting	K
Valves	
Globe, fully open	10
Angle, fully open	2
Gate, fully open	0.15
Gate, ¼ closed	0.26
Gate, ½ closed	2.1
Gate, ¾ closed	17
Swing check, forward flow	2
Swing check, backward flow	Infinity
180° return bends	
Flanged	0.2
Threaded	1.5
Elbows	
Regular 90°, flanged	0.3
Regular 90°, threaded	1.5
Long radius 90°, flanged	0.2
Long radius 90°, threaded	0.7
Long radius 45°, threaded	0.2
Regular 45°, threaded	0.4
Tees	
Line flow, flanged	0.2
Line flow, threaded	0.9
Branch flow, flanged	1.0
Branch flow, threaded	2.0

Table 11-2 Hazen–Williams coefficients (*C* has no units).

Material	C
Asbestos cement	140
Brass	130–140
Brick sewer	100
Cast-Iron	
New, unlined	130
10 yr. old	107–113
20 yr. old	89–100
30 yr. old	75–90
40 yr. old	64–83
Concrete/Concrete-lined	
Steel forms	140
Wooden forms	120
Centrifugally spun	135
Copper	130–140
Galvanized iron	120
Glass	140
Lead	130–140
Plastic	140–150
Steel	
Coal-tar enamel lined	145–150
New unlined	140–150
Riveted	110
Tin	130
Vitrif. clay (good condition)	110–140
Wood stave (avg. condition)	120

ing stream has a free or unconstrained surface that is open to the atmosphere. Flows in canals or in vented pipelines, which are not flowing full, are typical examples. The presence of the free water surface prevents transmission of pressure from one end of the conveyance channel to another as in fully flowing pipelines. Accordingly, in open channels, the only force that can cause flow is the force of gravity on the fluid. With steady, uniform flow under free discharge conditions, a progressive fall or decrease in the water surface elevation always occurs as the flow moves downstream.

In hydraulics, a pipe is any closed conduit that carries water under pressure. The filled conduit may be square, rec-

Table 11-3 Surface roughness.

Material	Surface roughness, e	
	Feet	Meters
PVC, plastic, glass	0.0	0.0
Commercial steel or wrought iron	1.5e–4	4.5e–5
Galvanized iron	5.0e–4	1.5e–4
Cast iron	8.5e–4	2.6e–4
Asphalted cast iron	4.0e–4	1.2e–4
Riveted steel	0.003 to 0.03	9.0e–4 to 9.0e–3
Drawn tubing	5.0e–6	1.5e–6
Wood stave	6.0e–4 to 3.0e–3	1.8e–4 to 9.0e–4
Concrete	0.001 to 0.01	3.0e–4 to 3.0e–3

tangular, or any other shape, but is usually round. If flow is occurring in a conduit but does not completely fill it, the flow is not considered pipe or closed-conduit flow, but is classified as open-channel flow.

Flow occurs in a pipeline when a pressure or head difference exists between ends. The rate or discharge that occurs depends mainly upon:

- the amount of pressure or head difference that exists from the inlet to the outlet;
- the friction or resistance to flow caused by pipe length, pipe roughness, bends, restrictions, changes in conduit shape and size, and the nature of the fluid flowing; and
- the cross-sectional area of the pipe.

Velocity head concept

To understand velocity head consider that an object, such as a rock, when dropped will rapidly gain speed as it falls. How rapidly that rock picks up speed has been measured. Tests show that an object falling 1 foot will reach a velocity of 8.02 feet per second (ft/s). An object that falls 4 feet will reach a velocity of 16.04 ft/s. An 8-ft drop causes the object to reach a velocity of 22.70 ft/s. This gain in speed or acceleration is caused by gravity, referred to as "g." The force which gravity (g) exerts on an object is equal to 32.2 feet per second per second (ft/s^2).

If water is stored in a tank and a small opening is made in the tank wall 1 foot below the water surface, the water will spout from the opening with a velocity of 8.02 ft/s. This velocity has the same magnitude that a freely falling rock attains after falling 1 foot. Similarly, openings located 4 feet and 8 feet below the water surface, result in the velocity of the spouting water being 16.04 and 22.68 ft/s, respectively. Thus, the velocity of water leaving an opening under a given head (h) is the same as the velocity that would be attained by an object falling that same distance. The equation that shows how velocity changes with head and defines velocity head is:

Velocity equals the square root of two times gravity times head.

In mathematical notation the equation would read:

$$V = \sqrt{2gh}$$

This may also be written in velocity head form as:

Head = velocity squared divided by two times gravity;

or

$$H = V^2/2g$$

Basic principles of water measurement

Most devices measure flow indirectly. Flow measuring devices are commonly classified into those that sense or measure velocity and those that measure pressure or head. Once head (or velocity) is measured, charts, tables, or equations are used to obtain the discharge.

Some water measuring devices that use measurement of head, h, or pressure, p, to determine discharge, Q, are

1. Weirs,
2. Flumes,
3. Orifices, and
4. Venturi meters.

Head or depth is commonly used for open-channel devices such as flumes and weirs. Tube-type flowmeters such as a venturi (Fig. 11-1) utilize pressure or head.

11-1 *A typical venturi.*

Pressure is the force per unit area that acts in every direction normal to containing or submerging object boundaries. In an open vertical tube inserted through and flush with the wall of a pipe under pressure, water will rise to a specific height. This specific height (h) occurs when the weight (W) of water in the tube balances the pressure on the wall opening area at the wall connection.

The volume of water in these tubes (piezometers) is designated h_a. Volume multiplied by the unit weight of water, equals weight. The pressure force on the tap connection area is designated p_a. The weight and pressure force are equal, and dividing both by the area (A) gives the unit pressure on the wall of the pipe in terms of head (h).

Correspondingly, head is pressure divided by unit weight of water (62.4 pounds per cubic foot). Since pressure is expressed in pounds per square inch (psi) or (lb/in^2) it can be converted to feet of water by multiplying the (lb/in^2) value by 2.31.

Example: 30 lb/in^2 pressure multiplied by 2.31 equals 69.3 feet of water.

Conversely, if the known factor is feet of water and you want to convert to pressure, simply divide by 2.31. Remember that your results will be in pounds per square inch.

Example: 69.3 feet of water divided by 2.31 equals 30 lb/in^2 of pressure.

Discharge

When the head principle is used, the discharge (Q) is computed from the equation:

$$Q = CLh^{1/2}$$

A coefficient, C, is included that accounts for simplifying assumptions and other deficiencies in deriving the equation. The coefficient can vary widely in nonstandard installations, but is well defined for standard installations or is constant over a specified range of discharge. One of the advantages of using computerized programs is that coefficients are automatically provided in the program.

The flow cross-sectional area A, does not appear directly in the equation, but an area can be extracted by rewriting this equation:

$$Q = CLh\, h^{1/2},$$

in which

$$A = Lh$$

In this form, C also contains a hidden square root of $2g$ which, when multiplied by $h^{1/2}$, is the theoretical velocity. This velocity does not need to be directly measured or sensed.

Because the weir equation computes velocity from a measuring head, a weir is classified as a head measuring device. Some devices that actually sample or sense velocities are

- Float and stopwatch,
- Current and propeller meters, and
- Vane deflection meters.

These devices generally do not measure the average velocity for an entire flow cross section. As a result, the relationship between sampled velocities, v, and the mean velocity, V, must be known as well as the flow section area, A, to which the mean velocity applies. The discharge, Q, sometimes called the flow rate, is the product of AV.

Discharge or rate of flow has units of volume divided by unit time. Thus, discharge can be accurately determined by measuring the time (t) required to fill a known volume (V_o).

Discharge equals known volume divided by measured time.

$$Q = V_o/t$$

Example: Very simply put, a tank known to hold 5 gallons that fills in 1 minute would yield a discharge rate of 5 gal/min.

Water measurement devices can be calibrated using very accurate volumetric tanks and clocks. More commonly, converting the weight of water per unit volume uses weight of water in the tanks. The weight of water per cubic foot, called unit weight or specific weight, is 62.4 lb/ft^3 at standard atmospheric conditions.

Continuity

The continuity concept is an important extension of the equation $Q = AV$. On the basis that water is incompressible and none is lost from a flowing system, then as the cross-sectional area changes, the velocity must adjust itself so that the values of Q or AV are constant:

$$Q = A_1V_1 = A_2V_2 = \ldots = A_nV_n,$$

where the subscript n denotes any number of arbitrarily selected positions along the flowing system. This principle, known as continuity, is especially useful in the analysis of tube-flow measurement devices, such as the venturi meter.

As already discussed, flow rate or discharge (Q), is the volume of water (in cubic feet) passing a flow section per unit of time, usually measured in cubic feet per second (ft^3/s). The distance (d_v) in feet that water will travel at a given velocity in a pipe of constant diameter is velocity (V) measured in feet per second (ft/s) multiplied by time (t) measured in seconds. Mathematically expressed:

$$d_v = Vt,$$

or

Distance equals velocity multiplied by measured time.

The volume (V_o) in cubic feet passing from the upstream end to the downstream end of this distance equals the distance (d_v) in feet, times area (A), in square feet, of the flow section. Thus,

$$V_o = AVt$$

To calculate the time rate of flow (discharge) in cubic feet per second, divide the right and left sides of the equation by t (time) in seconds. The results come back to:

$$Q = AV$$

Flow rate (discharge) Q equals the number of square feet of the flow section (A) times velocity (V).

Equation coefficients

The previous examples show that coefficients are used to correct for factors that are not fully accounted for in water measurement. These corrections are made using simplifying assumptions during derivations of equations. For the convenience of using a measured water head (h_1) instead of the more complex total head (H_1) the coefficient C_v is used because velocity head is often ignored in equations.

Orifices require an area correction to account for jet contraction in the orifice. The flow is forced to curve around and spring from the sharp edge, forming a contracted jet or vena contracta. The contracted area of flow, A_c, should be used in hydraulic relationships. Thus, the area, A_o, of the orifice must be corrected by a coefficient of contraction defined as:

$$C_c = A_c/A_o$$

Properly designed venturi meters and nozzles have no contraction, which makes C_c unity because of the smooth transitions that allow the water to flow parallel to the meter boundary surfaces. Ultimately, the actual discharge must be measured experimentally by calibration tests, and the theoretical discharge must be corrected.

A common misconception is that coefficients are constant. In fact a constant coefficient is only constant within a range of discharges, not with all discharges. Complying with structural and operational limits for standard devices will prevent measurement error caused by using coefficients outside of the proper ranges. Some water measuring devices cover wider ranges using variable coefficients of discharge by means of plots and tables of values with respect to head and geometry parameters. Coefficients also vary with measuring station head or pressure tap location. Therefore, users should make sure that the coefficients used match pressure or head measurement locations.

NOTE: Water measurement equations generally require use of some if not all of these coefficients to produce accurate results. Often, composite numerical coefficients are given that are the product of combinations of area (or a dimension factored from the area), acceleration of gravity, integration constants, and the correction coefficients. However, geometry dimensions and physical constants, such as acceleration of gravity, are better kept separate from the non-dimensional coefficients that account for the difference between theoretical and actual conditions. Otherwise, converting equations from English to metric units is more difficult.

The coefficient of discharge for venturi meters ranges from 0.9 to about unity in the turbulent flow range and varies with the diameter ratio of throat to pipe. The coefficient of discharge for orifices in pipes varies from 0.60 to 0.80 and varies with the diameter ratio. For flow nozzles in pipelines, the coefficient varies from 0.96 to 1.2 for turbulent flow and varies with the diameter ratio. ASME and ISO have detailed treatments of pipeline meter theory, coefficients, and instruction in their manuals.

Selection guidelines

Selection of a water measurement method can be a difficult, time-consuming process if you formally evaluate all the factors discussed for each measuring device. Of course, this difficulty is one reason that standardization of measurement devices within a district is so popular. However, useful devices are sometimes overlooked when similar devices are automatically selected. The purpose of this section is to provide some preliminary guidance on selection so that the number of choices can be narrowed down before a more thorough analysis of the trade-offs between alternatives is performed.

Site conditions for a water measurement device quickly narrow the list of possible choices, because most devices are only suitable under a limited number of channel or conduit conditions. Table 11-4 provides a list of the most commonly used measurement methods for each of several applications.

The process of narrowing down options might start with Table 11-4 examining the main methods to be considered. In narrowing down the options, different applications will place different weights on the selection criteria, so no universally correct selection exists. Finally, a preliminary design for several possible methods selected should be performed so that details on cost, hydraulics, operations, etc., can be more thoroughly examined.

Table 11-4 Application-based selection of water measurement devices.

- I. Open-channel conveyance system
  - A. Natural channels
    - 1. Rivers
      - a. Periodic current metering of a control section to establish
      - b. Stage-discharge relation
      - c. Broad-crested weirs
      - d. Long-throated flumes
      - e. Short-crested weirs
      - f. Acoustic velocity meters (AVM—transit time)
      - g. Acoustic Doppler velocity profiles
      - h. Float-velocity/area method
      - i. Slope-area method
    - 2. Intermediate-sized and small streams
      - a. Current metering/control section
      - b. Broad-crested weirs
      - c. Long-throated flumes
      - d. Short-crested weirs
      - e. Short-throated flumes
      - f. Acoustic velocity meters (AVM—transit time)
      - g. Float-velocity/area method
  - B. Regulated channels
      - a. Spillways
        - Gated
          - Sluice gates
          - Radial gates
      - b. Ungated
        - Broad-crested weirs (including special crest shapes, Ogee crest, etc.)
        - Short-crested weirs
      - c. Large canals
        - Control structures
        - Check gates
        - Sluice gates
        - Radial gates
        - Overshot gates
      - d. Other
        - Long-throated flumes
        - Broad-crested weirs
        - Short-throated flumes
        - Acoustic velocity meters
      - e. Small canals (including open-channel conduit flow)
        - Long-throated flumes

Table 11-4 Application-based selection of water measurement devices (Continued).

- Broad-crested weirs
- Short-throated flumes
  - Sharp-crested weirs
- Rated flow control structures (check gates, radial gates, sluice gates, overshot gates)
- Acoustic velocity meters
- Other
  - Float-velocity area methods

f. Farm turnouts
- Pipe turnouts (short inverted siphons, submerged culverts, etc.)
  - Metergates
  - Current meters
  - Weirs
  - Long-throated flumes
  - Short-throated flumes

g. Other
- Constant head orifice
- Rated sluice gates
- Movable weirs

h. Closed-conduit conveyance systems
- Large pipes
- Venturi meters
- Rated control gates (orifice)
- Acoustic velocity meters (transit time)

i. Small and intermediate-sized pipelines
- Venturi meters
- Orifices (in-line, end-cap, shunt meters, etc.)
- Propeller and turbine meters
- Magnetic meters
- Acoustic meters (transit-time and Doppler)
- Pitot meters
- Elbow meters
- Trajectory methods (e.g., California pipe method)
- Other commercially available meters

Example: Measuring the flow entering a small turnout ditch that serves an agricultural field. The ditch is trapezoidal, concrete-lined, and has a rectangular metal sluice gate that is opened by hand to divert flow into the ditch from a canal lateral. No power is available at the site. The ditch carries a flow of about 10 cubic feet per second (ft^3/s). Field survey measurements taken indicate a drop of approximately 0.75 ft in the water surface from the gate to the downstream channel. The irrigation flow transports fine sediment and plant debris. Water is diverted to the field on a two-week rotation for a period of about twenty-four hours. The measurement device used will establish a known flow rate through the headgate for crop yield management and water use accounting.

Typically, the water surface in the lateral remains constant during irrigation; therefore, a single measurement per irrigation will meet current needs. In the future, however, measurements made on a more frequent basis may be desired. The irrigator would like to install a device that costs less than $500.

- Table 11-4 identifies a number of devices that are typically used for farm turnouts. The site requires a device or method that can be used in an open channel. Therefore, common measurement devices given for this application are
  1. current meters,
  2. weirs,
  3. flumes, and
  4. rated sluice gates (headgates).
- Next, the advantages or disadvantages for each of these devices should be considered with respect to the measurement goals and the site conditions. Typically, only a few selection constraints are high priorities. The selection priorities for the example are likely to be
  1. meeting available head,
  2. cost,
  3. accuracy, and
  4. debris passage goals.

  Head loss is the highest priority because it is a physical constraint of the site that must be met to provide good measurement. Current meters provide the least head loss followed by long-throated flumes (including broad-crested weirs), short-throated flumes, and sharp-crested weirs. Sluice gates rate low in terms of head loss. However, for this application, the gate is part of the site and will not provide additional head loss. Based on the highest priority, current metering, a long-throated flume, or rating the headgate are good choices.

- Next, consider the cost of devices including:
  1. initial cost,
  2. data collection time, and
  3. maintenance.

  The headgate and long-throated flume are usually lower cost options than current metering, largely because of the time involved in data collection. Accuracy of measurement and debris passage favor a long-throated flume.

Now that the basics of fluid mechanics have been discussed, you should know that any of the measurements mentioned here can be automatically calculated using computer software. Software programs can be purchased and installed on your in-house system or you can utilize one of many interactive programs located on the World Wide Web. These sites require you to plug known data into well-labeled data fields and make a few simple selections as to what unknown(s) you wish to solve. Once the data has been entered, push the "calculate" button and all the answers appear, properly organized and ready to print-out. The reason for going through all of the numbers and equations in this section is to provide you with an understanding of the data needed for the computer programs.

Private water supplies

When one is confronted with a project dependent on a well (rather than the public system) as the source of water, several additional considerations come into play.

Well considerations

When the potable water source is a well, the well should be tested to make sure it can supply the volume of water at peak demand. If the volume available is not enough to satisfy peak demand, an extra large holding tank must be utilized.

Pump selection The basic type of pump to be used to force the water from the well to the structure must be decided upon, using two factors:

- The vertical distance from the pumping level of the water in the well to the holding tank (Fig. 11-2).
- The anticipated friction loss within the suction pipe transporting the water from the well to the suction opening of the well pump.

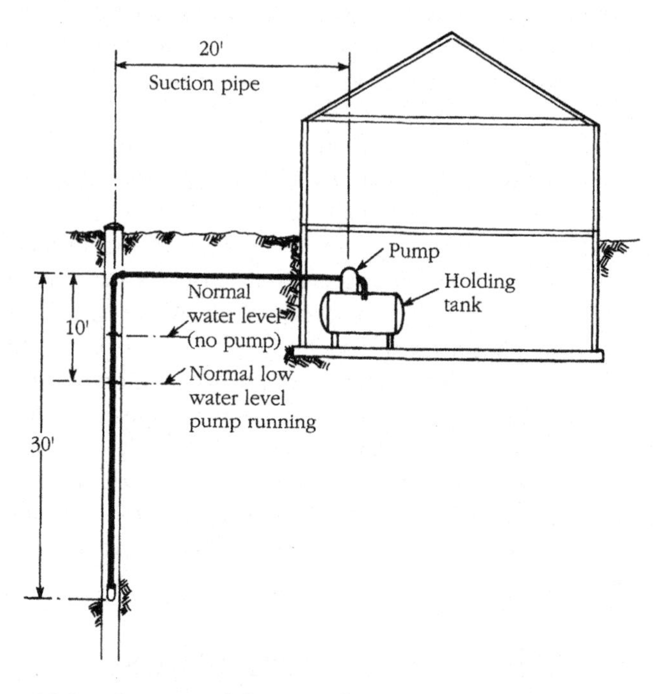

***11-2** The vertical distance from the pumping level of a well to the holding tank.*

The sum of these two factors equals the suction lift factor, which will determine if a shallow or deep well pump is to be used.

Calculating the anticipated friction loss within the suction pipe involves three elements: the length of the suction pipe, the size of the suction pipe, and the peak demands of the water system. Table 11-5 provides the friction loss factors based on gallons per hour water flow and pipe size.

Locate the peak demand in the first column of Table 11-5. Follow across the row to the number at the intersection of the column headed by the size of the pipe. This number is the fric-

Table 11-5 Suction pipe friction loss factors per 100 foot length.

PH	⅜" Pipe	½" Pipe	¾" Pipe	1" Pipe	1¼" Pipe	1½" Pipe	2" Pipe	2½" Pipe	3" Pipe	4" Pipe
60	4.30	1.86	0.26	0.11						
120	15.00	4.78	1.21	0.38						
180	31.80	10.00	2.50	0.77						
240	54.90	17.10	4.21	1.30	0.34					
300	83.50	25.80	6.32	1.93	0.51	0.24				
360		36.50	8.87	2.68	0.70	0.33	0.10			
420		48.70	11.80	3.56	0.93	0.44	0.13			
480		62.70	15.00	4.54	1.18	0.56	0.17			
540			18.80	5.65	1.46	0.69	0.21			
600			23.00	6.86	1.77	0.83	0.25	0.11	0.04	
720			32.60	9.62	2.48	1.16	0.34	0.15	0.05	
900			49.70	14.70	3.74	1.75	0.52	0.22	0.08	
1200			86.10	25.10	6.34	2.94	0.87	0.36	0.13	
1500				38.60	9.65	4.48	1.30	0.54	0.19	
1800				54.60	13.60	6.26	1.82	0.75	0.26	
2100				73.40	18.20	8.37	2.42	1.00	0.35	
2400				95.00	23.50	10.79	3.10	1.28	0.44	
2700					30.70	13.45	3.85	1.60	0.55	
3000					36.00	16.40	4.67	1.94	0.66	0.18
4200					68.80	31.30	8.86	3.63	1.22	0.35
5000						62.20	17.40	7.11	2.39	0.63
9000							38.00	15.40	5.14	1.32
2000							66.30	26.70	8.90	2.27
5000							90.70	42.80	14.10	3.60
3000								58.50	19.20	4.89

tion factor for that pipe size at the level of peak demand per 100 feet of pipe length. For pipe lengths shorter than or greater than 100 feet, divide the actual length by 100; this results in a percentage by which the friction factor should be multiplied to adjust for the shorter pipe length. For example, for a 65-foot length of pipe and a peak demand of 60 gph, the calculations would be as follows:

$$\frac{65}{100} = 0.65$$

$$0.65 \times 4.30 = 2.79$$

If the length of the pipe in the example were 120 feet instead, the calculations would be:

$$\frac{120}{100} = 1.20$$

$$1.20 \times 4.30 = 5.16$$

If the total suction lift (vertical distance from pumping level of water to holding tank, plus friction loss) is less than 25 feet, a shallow well pump is indicated. Should the total suction lift be equal to or greater than 25 feet, a deep well pump will be necessary.

Operating pressures The operation of the pump is generally dictated by pressure switches inside the holding tank, which turn the pump on and off when pre-set pressure levels are reached in the tank. Although the normal operating pressure settings within the holding tank are 20 pounds per square inch (psi) for cut-in and 40 psi for cut-out, these pressures are affected by three factors:

- Friction loss within the discharge piping
- Service pressure required at the outlet
- Vertical distance from the holding tank (or pump) to the most remote outlet.

The friction loss of the discharge piping can be found using the figures in Table 11-5 and multiplying the appropriate friction loss factor by 0.434 to convert from feet to pounds. The vertical distance of the most remote outlet must also be converted to pounds of pressure per square inch—accomplished by dividing the distance in feet by 2.31.

If one or more of these three factors reflect above average conditions, a pump must be selected that cannot only supply the volume of water required but also deliver it at a higher pressure. The cut-in switch inside the tank must also be set to operate a the higher pressure setting to compensate for the anomaly as well.

Water supply problems

In troubleshooting water supply problems, mathematics becomes very helpful. Some of the most common water supply problems are discussed in this section.

Low outlet pressure The water pressure at any outlet is always equal to the tank pressure less the sum of any pressure losses occurring between the tank and that outlet. With pressure losses in the line, low outlet pressure can result.

One of the causes of pressure loss is the difference in height from the holding tank to the remote outlet. For every 2.3 feet of vertical run there is a drop of 1 pound per square inch of pressure.

$$2.3 \text{ feet} = 1 \text{ psi}$$

A height difference of 30 feet from the tank to the outlet results in a drop in pressure of 13 psi (30 / 2.3 = 13).

The real source of low-pressure problems is at the pump. Low water pressure is indicative of insufficient pump capacity at higher pressures. It is therefore necessary to select a pump that delivers a rate of flow into the tank, at a pressure great enough to compensate for any pressure losses experienced within the system.

To ensure the proper pump capacity, calculate the vertical distance from the tank to the highest outlet. Using the formula above, convert the distance in feet to pressure loss in pounds per square inch. To this figure, add the friction loss (from Table 11-5) corresponding to the length and size of pipe supplying that outlet. Once this figure is known, add an additional 5 or 10 psi to ensure adequate discharge pressure.

Bear in mind that the capacity of jet pumps and submersible pumps have an inverse relationship with the pressure inside the tank—the pump capacity will increase as the pressure in the tank decreases and will decrease as the pressure in the tank increases. The maximum capacity output will correspond to low tank pressure and will decrease as the tank pressure increases until there is actually no water being delivered from the pump.

Having calculated the gallon per hour demand on the system, the capacity of a pump may seem inadequate to furnish the water at required pressure. However, before selecting a larger pump, consider the total suction lift and how it is affected by the size of the suction pipe stipulated. An inverse relationship exists between the size of the suction pipe and the suction lift factor; the smaller the pipe, the greater the friction loss per foot of pipe run. For example, for a situation requiring 720 gph, having a 50-foot run of ¾-inch suction pipe and 10 feet of vertical height results in a total suction lift (vertical height plus friction loss) of 26.3 feet, 16.3 feet of which are due to friction loss. Under these circumstances, a given size pump may have its capacity reduced to less than 700 gph. However, the capacity of the same pump will increase to over 800 gph by replacing the ¾-inch suction pipe with 1-inch suction line due to the decrease in the total suction lift from 26.3 feet to 14.81 feet. The 11.49 feet of difference in the total suction lift is entirely attributable to the decrease in friction loss realized by using a larger suction pipe.

Math & pipe measurements

The use of mathematics is essential to properly measure the pipe used in a plumbing job to ensure the proper fit, angle, and junction of the plumbing lines.

Pipe material

The various materials used to manufacture the pipe for water supply lines include: copper, steel, plastic, and brass.

- *Copper.* Copper has long been the plumbing line of choice. Still widely used and very popular today, copper tubing is available in two forms; rigid and flexible, with letter designations for various applications. Rigid copper tubing is supplied in straight runs of different lengths (normally 10 or 20 feet), while flexible copper is most commonly seen in level wound or pancaked coils. Because of its ductility, copper is popular in situations where straight runs are not the norm. Being a "soft" metal, copper is easily adaptable to areas requiring the pipe to bend in order to make a connection.
- *Plastic.* Several types of plastic resin are used today to form plastic pipe for plumbing purposes. Lighter in weight and less expensive than copper, plastic has become widely used in both new construction and retrofit.
- *Steel.* Galvanized steel pipe has all but vanished from new construction because of the labor-intense installation process. Each length of pipe must be cut to fit, and the ends are threaded to fit screw fittings.

- *Brass.* As with galvanized steel, brass is hardly ever used today. Popular for a period of time in the past, the expense of brass as plumbing pipe is prohibitive.

When replacing plumbing lines in an existing structure, the existing type of pipe should be matched whenever possible.

Measuring pipe

Copper, plastic, and galvanized steel pipe are most commonly measured using a method referred to as face-to-face—probably the least complicated manner in which to measure pipe. The exact distance from the face of the fitting on one side of the pipe to the face of the fitting to which the pipe will connect is measured first. To this figure must be added the amount of pipe that will be secured inside the fittings. Remember, regardless of what type of pipe is being used, there will be an additional length of pipe on each end that will be either soldered, screwed, or chemically welded inside of the fittings. For example, if ½-inch type L copper tubing is being used and the face-to-face measurement is 8 feet 6 inches, generally ½ inch should be allowed on each end to account for the tubing which extends inside the fittings. The total length of the ½ L water tube should be 8 feet 7 inches.

An alternate method to the face-to-face is the center-to-center method. The equation for the center-to-center method, as illustrated in Fig. 11-3 is as follows:

$$D = A - 2B + 2C$$

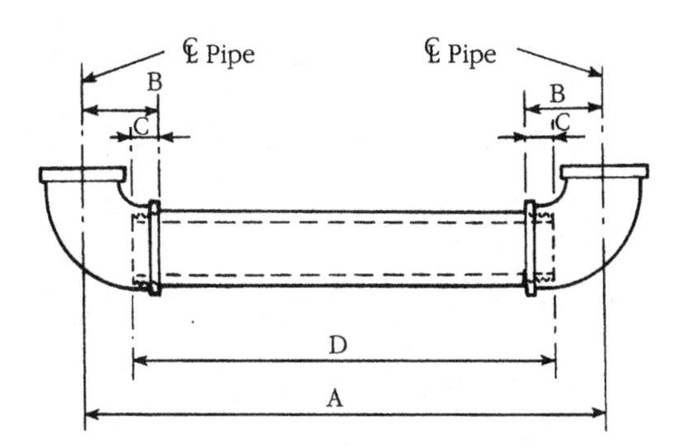

11-3 *Example of center-to-center pipe measurements.*

where D stands for distance (the value being calculated). The A represents the measured distance from the exact center of one fitting to the exact center of the fitting on the opposing end of the pipe. The B represents the distance from the face of one fitting to the center of that same fitting. Because the pipe will not extend into the fitting as far as the center of that fitting, the length of B must be subtracted from both ends of the pipe, which is why the formula calls for the subtraction of $2B$. The C in the formula represents the actual length of pipe that will extend into the fittings. Here again, since this occurs on both ends of the pipe, the formula stipulates the addition of $2C$. Table 11-6 provides center-to-face distances (the B values in the formula) for 45-degree and 90-degree elbow fittings as well, as for tee fittings and cross fittings as they correspond to common pipe sizes.

Table 11-6 Center-to-face distances of common fittings as they relate to different pipe sizes.

When used with 45 degree elbow fittings and various size pipes		***When used with 90 degree elbow, tee or cross fittings and various size pipes***	
Pipe size	**Center-to-face distance**	**Pipe size**	**Center-to-face distance**
1/8 inch	3/4 inch	1/8 inch	11/16 inch
1/4 inch	3/4 inch	1/4 inch	13/16 inch
3/8 inch	13/16 inch	3/8 inch	15/16 inch
1/2 inch	7/8 inch	1/2 inch	1 1/8 inch
3/4 inch	1 inch	3/4 inch	15/16 inch
1 inch	1 1/8 inch	1 inch	1 1/2 inch
1 1/4 inch	15/16 inch	1 1/4 inch	1 3/4 inch
1 1/2 inch	17/16 inch	1 1/2 inch	11 5/16 inch
2 inch	1 11/16 inch	2 inch	2 1/4 inches
2 1/2 inch	1 15/16 inch	2 1/2 inch	2 11/16 inches
3 inch	2 3/16 inches	3 inch	3 1/16 inches
3 1/2 inch	2 19/32 inches	3 1/2 inch	3 27/32 inches
4 inch	2 5/8 inches	4 inch	3 13/16 inches
5 inch	3 1/16 inches	5 inch	4 1/2 inches
6 inch	13 15/22 inches	6 inch	5 1/2 inches

Multiply the center-to-face value by 2 when fittings will be used at both ends of the pipe.

Yet another formula is employed in circumstances requiring offsetting pipes in order to make the needed connections. Referring to Fig. 11-4, the dotted lines (*A* and *B*) when added to the solid line (*C*) form the triangle created by the offset. The trigonometric equation used to determine the length of pipe connecting the two fittings would be:

$$AC = \text{the square root of } (AB^2 + BC^2)$$

where *AC* is the length of the pipe (the value being calculated), and *AB* and *BC* equal the offsets.

Again, using Fig. 11-4, the distance between pipe *E* and pipe *F* (the two water lines being connected) is the offset represented by line *AB*. For demonstration purposes, assume this distance is equal to 30 inches. Since 45 degree elbows are shown in this illustration, a mathematical constant comes into play; anytime two 45-degree fittings are used with such an offset, both offsets will be of equal length. By substituting the assumed values in place of *AB* and *BC*, the equation now reads:

$$AC = \text{the square root of } (30^2 + 30^2)$$

Performing the calculations for the equation results in the following:

$$AC = \text{the square root of } 1800 = 42.43 \text{ inches}$$

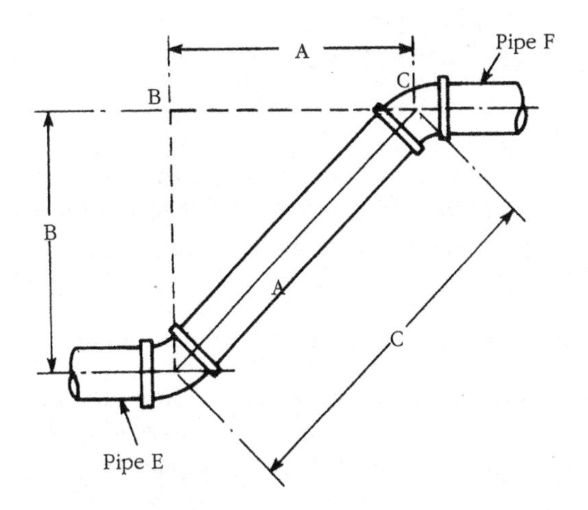

11-4 *Determining distances for offset pipe connections.*

Using this method results in a center-to-center measurement of the connecting pipe. Consequently, the center-to-face measurements (the amount of pipe extending inside the fittings at both ends) must be determined from Table 11-6 and subtracted from the calculated length of the connecting pipe. After subtracting the center-to-face dimensions, the actual length of pipe which will extend into the fittings at either end must be added to the length of the pipe as discussed earlier.

Not all elbows are 45-degree elbows, to simplify the calculations Table 11-7 lists multipliers (constants) for the various angles of common elbow fittings. Multiplying the value of the *AB* offset by the appropriate constant under the heading "*AC*" from Table 11-7 will yield the distance of the offset *AC*.

$$AC = \text{offset } AB \times \text{multiplier from Table 11-7}$$

Table 11-7 Multipliers for elbow fittings of various degree dimensions.

Elbow angle	AC	AB
60 degrees	1.15	0.58
45 degrees	1.41	1.00
30 degrees	2.00	1.73
22½ degrees	2.61	2.41
11¼ degrees	5.12	5.02
5⅝ degrees	10.20	10.15

Also listed in Table 11-7 are constants (multipliers) used to determine the distance of offset *BC*. Multiplying the value of *AB* by the appropriate constant in the column headed by "AB" yields the length of offset *BC*.

$$BC = \text{offset } AB \times \text{multiplier from Table 11-7}$$

For example, using the earlier values with *AB* equal to 30 inches, assume the fittings are 60-degree instead of 45-degree elbows. Referring to Table 11-7, the multiplier for a 60-degree elbow in column *AC* is 1.15. Substituting the values into the equation results in the equation reading:

$$AC = 30 \times 1.15$$
$$AC = 34.5$$

To determine the length of *BC*, locate the multiplier in the column under "AB", which corresponds to a 60-degree elbow (0.58) and substitute the values into the equation for *BC* given previously.

$$BC = 30 \times 0.58$$
$$BC = 17.40$$

Planning pipe layouts

The clear air space inside walls or between floors and ceilings must be large enough to accommodate the pipes and fittings which must run through them.

Since the majority of plumbing today is done with either copper or plastic, the available space in these areas is less critical than it used to be when galvanized steel was popular. A 2-×-4 stud wall will generally afford 3⅝ inches of clearance, enough to allow the use of 3-inch copper or thin walled plastic lines. 2-×-6 framing will normally have 5½ inches of clearance, while 2-×-8 framing generally allows 7½ inches of space.

Even spaces as tight as 1¾ inches can accommodate a 1½-inch line of copper or plastic. This is possible due to the fact that the pipe can be cut to required length and the fittings can be located elsewhere. With this type of versatility, even a 3-inch pipe can be run through a space as small as 3¼ inches. Table 11-8 presents the minimum space requirements for drainage/waste/vent (DWV) pipes.

Table 11-8 Space requirements for laying out DWV lines.

Size of pipe	Plastic	Copper
1½ inch	3 inches	2 inches
2 inch	3 inches	3 inches
3 inch	4½ inches	3½ inches
4 inch	5½ inches	4½ inches

Planning a bathroom

No longer are bathrooms the tiny, utilitarian rooms of years gone by. With the advent of platform tubs, sunken tubs, hot tubs, whirlpools, saunas, and other luxury upgrades, bathrooms have taken on more importance for the homeowner, which requires more planning for the contractor.

Some of these newly popular items present special problems that must be planned for by the contractor(s). A sunken tub flush with the level of the floor is a good example. Providing that sufficient space under the floor exists for the tub and its drain fittings, most tubs of this nature require more fixtures and lines than normal tub or tub/showers do.

Types of fixtures

Most tubs are built from one of three common materials:

- *Fiber-glass reinforced plastic* is highly malleable and can be molded into many shapes. Most tubs made from it weigh between 60 and 70 pounds, are sturdily reinforced, and are warmer to the touch than others. Although this material doesn't chip easily, if at all, abrasive cleaners will damage the surface. In addition, installing the tub is sometimes a difficult job—all edges and stress points must be fully supported.
- *Molded cast iron* (with a porcelain enamel surface) makes an extraordinarily durable tub. It also makes an extraordinarily heavy tub—from 350 to 500 pounds. Like fiber glass, cast iron keeps water warm for a long time.
- *Formed steel tubs* (also with a porcelain enamel surface) are less expensive than cast iron and not nearly so heavy. They usually weigh between 120 and 125 pounds. Also, water cools down more quickly in a formed steel tub than in one made from fiber-glass reinforced plastic or cast iron, and steel tubs are prone to chipping and denting.

A typical tub is 5 feet long, but are also available as short as 4½ feet and as long as 8 feet. Receptor tubs are approximately 36 to 38 inches long, 39 to 42 inches wide, and about 12 inches high. They are most suitable for shower installations, but because of their lower height, are also convenient for bathing children and others who need assistance. Ordinary shower stalls range in size from a compact 32 inches square, up to 48 inches. Generally, a comfortable minimum is 36 inches.

Lavatories are available in a variety of materials, too.

- *Porcelainized cast iron* is extraordinarily durable, but heavy, which means it needs a sturdy support system.
- *Enameled steel* doesn't wear as well but is especially good when remodeling because it's light enough to move into place easily.
- *Stainless steel* is light, durable, and unaffected by household chemicals, but the steel tends to collect spots from hard water and soap residue.
- *Vitreous china* is easy to clean and has a lustrous surface, but can crack or chip when struck with a heavy object.
- *Fiberglass reinforced plastic* can be molded into novel shapes, but it doesn't hold a shine as well as other surfaces.
- *Marbleized china* has all the sterling qualities of the best natural china.
- *Simulated* or *cultured marble* is handsome, but abrasive cleaners may spoil the finish.

Just as lavatories vary in shapes, sizes, and materials, they are mounted in a number of ways; basically they hang from the wall, stand on their own pedestals, or rest in vanity cabinets.

- *Self-rimming*, or *surface-mounted*, lavatories feature a ridge around the bowl that fits over the counter top to form a tight seal. The ridge also prevents water from splashing onto the counter.
- A *flush-mounted* lavatory is recessed into the counter top with a tight-fitting metal rim around the bowl. The rim comes in different finishes to match the faucet. With this model, water frequently escapes onto the counter, making the rim joint hard to clean.
- A *recessed* lavatory fits under the counter top (which has a cutout in the surface) and also is difficult to clean around the edge.
- *One-piece integral* lavatories are molded with no joint or separation between the bowl and counter top.

Toilets are also a varied lot:

- *Integral units.* Some toilets feature tank and bowl combined into one low-to-the-floor piece. Ordinarily, their flushing sound is quieter than standard styles with the water tank above the bowl.
- *Bowl shapes.* Round bowls are most common, but some toilets have elongated bowls that are 2 inches longer, front to back, than their conventional cousins.
- *The space-saver.* A triangular toilet that fits neatly into a 2-×-2-foot corner.
- *Special features.* Some toilet tanks are insulated to stop water from condensing on the surface. Another model has an internal ventilating system designed to prevent odors. Still other toilets—those installed in the basement, for example—are built to flush upward.
- *For the disabled.* Toilet seats with arms projecting from them help in standing up and sitting down. One model is 18 inches high, compared to the normal 14 inches; the extra height makes it easier to use.
- *Interior design.* Toilets not only look different on the outside, they also work differently on the inside. The most common design, known as a *reverse trap,* flushes through an outlet in the back. Reverse-trap toilets are efficient and relatively quiet.
- *Siphon-jet* toilets improve on the reverse-trap design. This one also flushes through the back, but has a larger outlet that drains faster and more efficiently. Siphon-jet toilets are nearly silent, and cost considerably more than reverse-trap types.

Supply & DWV lines

The DWV system normally uses 4-inch sewer lines; 3- or 4-inch main drains and vents, serving toilets and groups of fixtures; 3- or 4-inch toilet drains; 2-inch shower drains; 1-inch drains for sinks and tubs; 1-inch lavatory drains. The wet wall is usually a few inches thicker than other walls to accommodate the 3- or 4-inch-diameter stack required in residential construction. Hot and cold supply lines may be in the wet wall. Always plan piping layouts to take the most direct route, avoiding obstacles wherever possible. DWV piping takes precedence over water-supply piping because it is bulkier and costlier. The smaller water pipes usually can be routed along with the DWV lines.

Pipes in the floor should run parallel to the joists where possible. Pipes in the attic can easily run across the tops of attic-floor joists while pipes below the first floor can be run under first-floor joists. Joists between floors must be notched or have holes cut in them for pipes running across them.

Septic tank systems

Septic systems are designed to safely dispose of biological sanitary waste on-site. The purpose of a home's subsurface sewage disposal system (septic system) is to dispose of the water generated by the occupants. This system needs to perform in such a manner that the soils on the property can disperse it without causing an adverse effect on ground water and in turn on public health and the environment.

Components of a system

As seen in Fig. 11-5, a septic system normally consists of the following four components:

- A sewer line that connects the home's plumbing to the septic tank.
- A septic tank that allows for the settling of soils and provides the initial treatment of the sewage.
- A distribution system that directs the flow of effluent from the septic tank to the leaching system in such a manner as to ensure full utilization of the system.
- A leaching or absorption system, which disperses the sewage effluent into the surrounding natural soils.

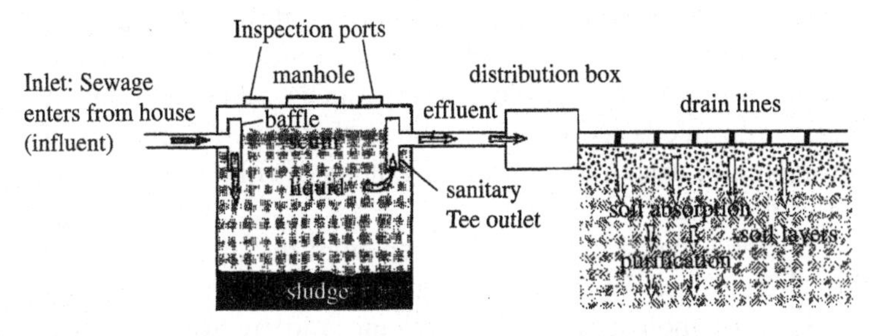

11-5 *The four typical components of a septic system.*

The sewer line Since sewage flows to the septic tank through the house sewer, this line must have the proper slope. The slope should not be so steep that the liquids run away from the solids and so flat that the solids settle out in the sewer pipe. A grade of from one to two inches per eight feet of run (a slope of one to two percent) is a rule of thumb. A one-percent slope is a one-foot drop in one hundred feet of pipe. Consequently, a fifty-foot run would require a six-inch drop, while a three-hundred-foot run would require a three-foot drop.

The house sewer should not have any low spots where liquid can pool. In freezing climates, these low spots are the places where sewer pipe freezing happens. A sagging sewer pipe and a dripping faucet are usually sure signs of a frozen house sewer in the northern climates. The sewer pipe should also be smooth on the inside so that sewage will not catch and dam. Rough spots at pipe joints present the possibility of a continuing problem of toilet paper getting hung up and plugging the pipe every so often.

The sewer pipe from the house becomes the inlet pipe to the septic tank. The bottom (invert) of this inlet pipe should be two to three inches higher than the invert of the outlet pipe of the septic tank. As the sewage reaches the tank, it drops into the liquid in the tank with a downward flow. This drop tends to move the sewage into the depth of the tank. Most states specify an inlet device, either a baffle or a sanitary tee. The purpose of the inlet device is to prevent the floating solids, called the scum layer, from building up and plugging the end of the sewer pipe. Make sure to check local codes to ensure compliance.

The tank The septic tank is the essential first part of an on-site sewage treatment system. Raw sewage flows into the tank from the house sewer. The solids separate from the liquid and stay in the tank. The liquid flowing out is called effluent. A septic tank needs to be watertight so that when 5 gallons of sewage enter from a toilet flush, 5 gallons of effluent must flow out of the tank.

Bacteria that do not use oxygen from the air grow inside the tank. These bacteria are called anaerobic and the by-products of their activity are methane and hydrogen sulfide gas, plus other substances having odor. Hence, the word "septic" has been applied to this tank. However, the septic tank is actually a settling tank where bacterial action takes place. The end products of the bacterial action are mainly water; gases; and undigested material, called sludge, that sinks to the bottom of the tank, and scum, that floats to the top of the tank. The septic tank contains baffles that prevent any scum that floats to the surface or sludge that settles to the bottom from passing out of the tank. Gases that are generated vent to the atmosphere via the plumbing vent system.

Sewage solids (sludge) are stored at the bottom of the tank while bacteria decompose them reducing their volume. The volume of the solids never reaches zero, so it must be pumped out of the tank periodically when the volume becomes too great.

The distribution system Figure 11-6 shows perforated pipes inside a distribution box. From the distribution box the flow of effluent is directed to the leaching system in such a manner as to ensure full utilization of the system. Most systems are "gravity" systems, meaning the flow runs through piping and distribution boxes without the assistance of any mechanical device, such as a pump or siphon.

The leaching (absorption) system A leaching system, also known as the absorption field or drainfield, disperses the sewage effluent into the surrounding natural soils. The soil also acts as a filter (Fig. 11-7) to remove any small amounts of solids that get carried along with the liquid. There are many types of leaching systems. The specific type utilized on a particular property is usually dependent on the soil conditions that exist on the site.

Soils vary in their ability to absorb and treat wastewater. Well-drained, medium-textured soils such as loam are best. Coarse gravel or sandy soils allow wastewater to flow too quickly for treatment. In fine clay or compacted soils, water moves too slowly. Soil microbes need oxygen to digest wastes quickly. If the air spaces between soil particles remain filled with water, the lack of oxygen prevents the rapid breakdown of wastes by aerobic (oxygen-requiring) soil microbes.

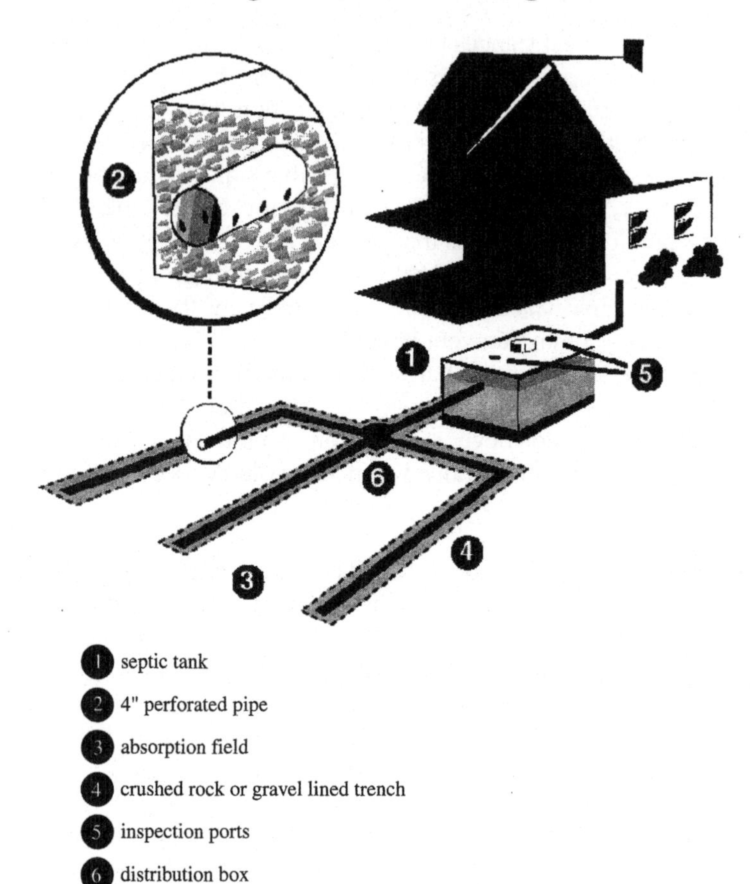

11-6 *From the distribution box the flow of effluent is directed to the leaching system.*

Anaerobic soil microbes (those that live in the absence of oxygen) digest wastes slowly and give off putrid, smelly gases characteristic of a failing septic system. Anaerobic conditions occur when soils are poorly drained, groundwater levels are high, surface runoff saturates the drainfield, or excessive amounts of water are used in the household.

Good wastewater treatment depends on good dispersal of wastewater over the drainfield. In a conventional, gravity-fed distribution system, the distribution pipes are often laid out in a fork-shaped pattern joined by a distribution box. Leveling de-

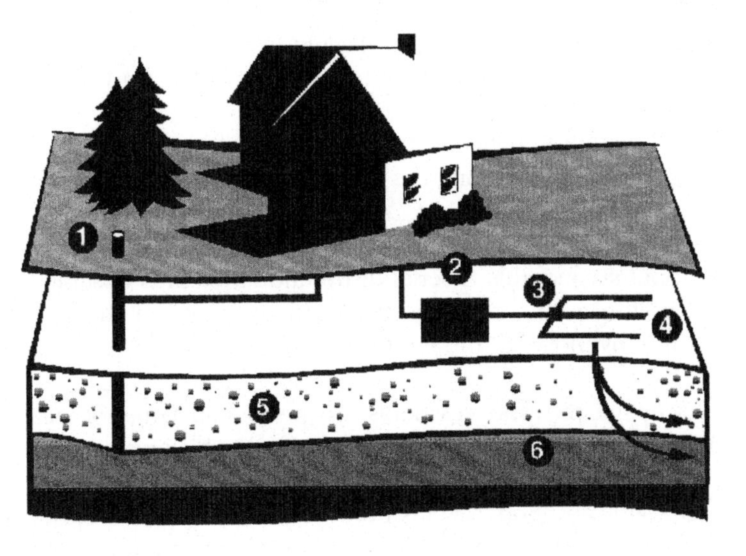

1 drinking water well

2 septic tank

3 distribution box

4 absorption field

5 soil absorption (unsaturated zone)

6 groundwater (saturated zone)

11-7 *Soil also acts as a filter.*

vices on the distribution box help ensure an even flow of wastewater to every trench. Often, however, certain trenches or low points in the distribution system receive more effluent than others do.

A dosing or enhanced-flow system has a pump or siphon to improve the distribution of effluent. Periodically pumping a certain volume of effluent to wet the entire drainfield area and then allowing the soil to drain between doses provides a period of aeration, which helps microorganisms in the soil digest the wastes.

In a pressure distribution system, effluent is pumped directly through small-diameter pipes and is not sent through a distribution box. Wastewater is evenly distributed throughout the entire drainfield, promoting better treatment of wastewater and system longevity.

Alternating trenches is another means of providing a period of aeration. Adjusting the outlet levels, or using a plug or valve in the distribution box, allows effluent to flow into only some of the trenches while other trenches are allowed to rest for about six months. A serial distribution system is designed so that the trenches are used in sequence; when the first trench is overloaded, the wastewater overflows into the next trench.

Seepage pits and cesspools are perforated tanks or pits lined with concrete blocks or bricks through which wastewater can seep into the ground. They are usually less effective than other soil absorption systems because they are located closer to the water table than trenches and often lack sufficient soil surface area for good wastewater treatment. Without a septic tank for pretreatment, a cesspool has the added problem of sludge accumulation in the pit. Cesspools are banned in some states.

Locating the components

Septic tanks are buried in the ground, usually a minimum of 10 feet from the house. The top of the tank is usually about one foot below the soil surface so it can be periodically opened for inspection and pumping. The distribution box is much smaller than the septic tank and is usual found about 20 feet from the house. It too is usually only about one foot below the ground.

From the distribution box, several pipes direct liquid to a series of pipes in trenches called laterals. The pipes in the trenches have holes in them to allow the liquid to be evenly distributed within the trench. To keep the pipes from being blocked with soil and to provide a space for water to be stored while it is being absorbed by the soil, the pipes are laid in a bed of crushed stone.

Above the stone is a soil filter (usually one or two layers of what is called untreated building paper). Above the soil filter is top soil in which grass is planted.

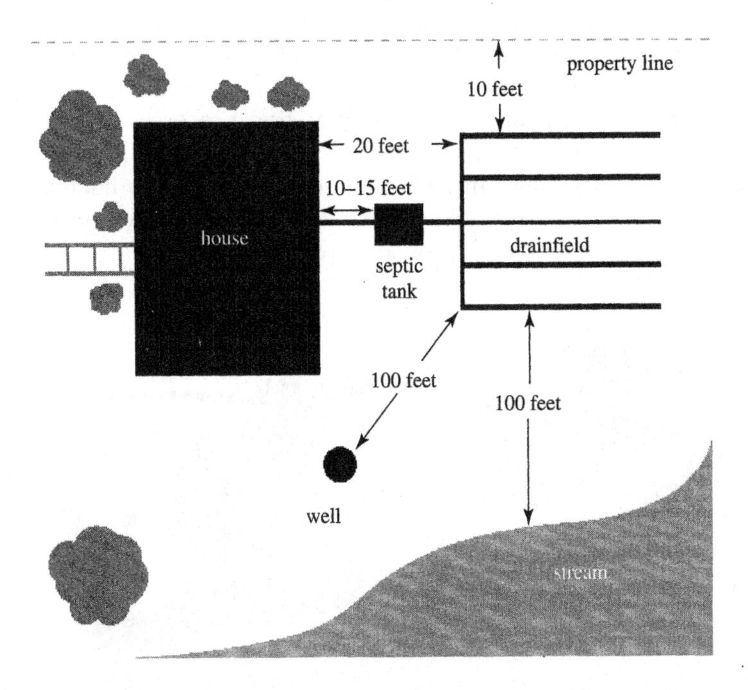

11-8 A typical septic system layout.

Where the components should not be

Figure 11-8 demonstrates a typical septic system layout. A septic system must be located a certain distance away from drinking water wells, streams, lakes, and houses. These distances are referred to as horizontal separation distances. Actual horizontal separation distances have been established and are specified in local regulations.

If there are wells, either on the property or adjacent to the property, the leach field must typically be a minimum of 100 feet from the location of the well. In some areas, the well is not allowed to be downslope from the leach field. If there is a stream or pond, the leach field must also be a minimum of 100 feet from the mean high water mark. Normally, no part of the system should be within 10 feet of a property line. In some areas and in unusual conditions, minimum distances may be greater than those noted here. In addition, no part of the system should be under a porch or driveway and you should not drive heavy vehicles (including automobiles) over the system lest the system be damaged.

In order to maintain aerobic digestion processes and remove contaminants effectively, the absorption field must be adequately separated from the groundwater or other limiting layer (Fig. 11-9). This is known as the vertical separation distance and is also specified by local regulations.

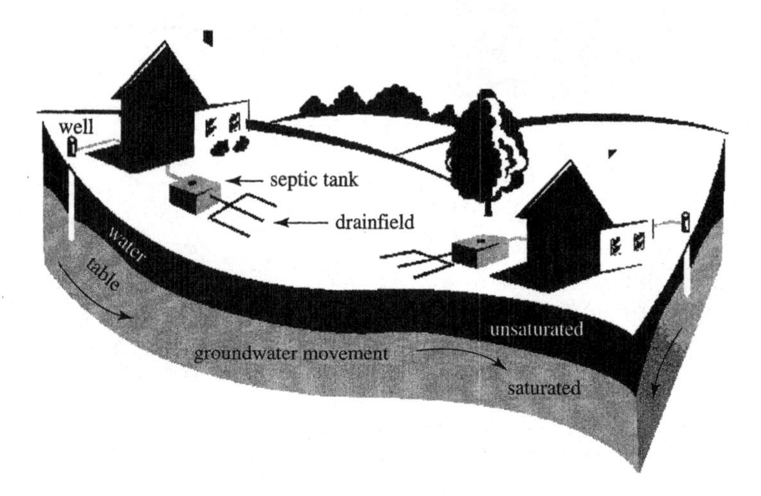

11-9 *Horizontal and vertical separation distances.*

Sizing a septic system

Septic systems are designed to dispose of household biological waste. The amount of waste to be handled depends on a number of factors. Among these are the number of people living in the house and their lifestyle. After many years of experience, a major guideline in determining the size and capability of a septic system has been correlated to the number of bedrooms in a house. The number of bedrooms typically determines the number of people generating waste and hence the amount of waste that must be handled.

Septic tanks are sized according to the amount of liquid waste they must process. The number of bedrooms in the home has been used as a rule of thumb for estimating water usage. For example: in New York State the minimum size tank that can be installed now is 1000 gallons for a 1, 2, or 3 bedroom house. For each bedroom after the first three, another 250 gallons needs to be added to the size of the tank. If a garbage grinder is in the kitchen sink, it counts as one additional bedroom. Table 11-9 provides some typical residential water usage for people, appliances, and activities.

Table 11-9 Home and outdoor living water requirements.

Use	Flow rate (gpm)	Total use (U.S. gallons)
Adult or child		50–100/day
Baby		100/day
Automatic washer	5	30–50/load
Dishwasher	2	7–15/load
Garbage disposal	3	4–6/day
Kitchen sink	3	2–4/use
Shower or tub	5	25–60/use
Toilet flush	3	4–7/use
Bathroom lavatory sink	2	1–2/use
Water softener regeneration	5	50/100/cycle
Backwash filters	10	100–200/backwashing
Outside hose faucet	5	?

Water flow restricting valves and shower heads can reduce flow and water use by up to 50%.

The septic tank should have adequate capacity to treat all the wastewater generated in the house, even at times of peak use. The system must be designed for maximum occupancy. Remember, along with the number of bedrooms, there may be more than one occupant for each bedroom. Water usage in the United States ranges from 50 to 100 gallons per day (gpd) per person. Knowing the number of occupants allows a check on the calculations of water usage.

Septic tanks should be large enough to hold two days worth of wastewater. (Two days is generally long enough to allow solids to settle out by gravity.) Typically, a new three-bedroom home is equipped with a 1000-gallon tank. A two-compartment tank (Fig. 11-10) or a second tank in series can improve sludge and scum removal and help prevent drainfield clogging.

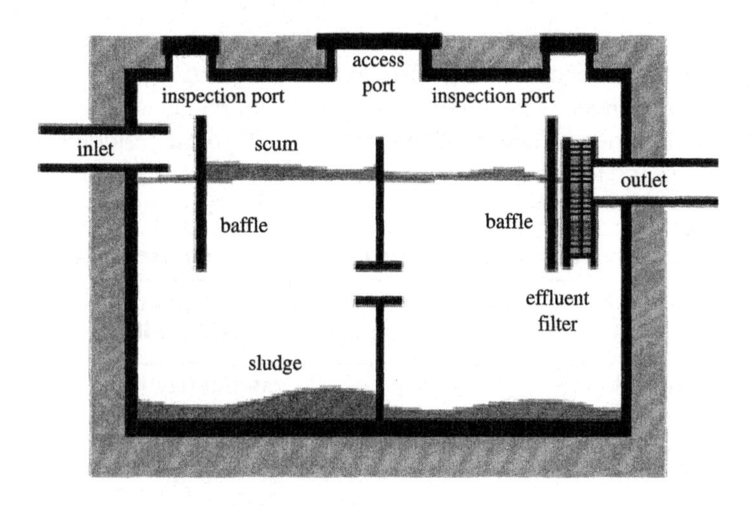

11-10 *A typical two-compartment tank.*

Sizing the leach field

Determining the required size of a leach field is a bit more complicated. The first thing to consider is the nature of the soil in which the leach field is to be constructed. Because water has to be absorbed in the soil, you need to know how fast it will be absorbed. This is called the percolation rate and is expressed as the time it takes for water in a test hole to decrease in level by one inch (minutes/inch). Another factor that must be known is the type of soil and how, or if, seasonal changes in the natural level of groundwater will interfere with the satisfactory operation of the system.

The primary consideration of planning a septic system is the absorption capacity of the ground. The total square feet of drainage trench required for the system will depend on the absorption capacity of the ground through which the trench runs.

Ground absorption is generally based on the results of a percolation test. This test involves a 1-foot square hole excavated to a depth equal to the estimated depth of the drainage trench (18 to 24 inches). With the hole filled with water, the perk rate or seeping time will equal the amount of time (in minutes) required for the water level to drop 6 inches.

Seasonal groundwater must be more than four feet from the bottom of the leach field trenches. If the soil percolates very fast (less than one minute per inch) or very slow (greater than 60 minutes per inch) it will not be possible to install a standard leach field in the existing soil.

The next step is to determine the amount of water that has to be absorbed each day. As with the septic tank sizing, there are also rules of thumb that can be used for finding out how much water must be absorbed each day for each bedroom in the house (expressed as gallons per day per bedroom). For older houses (built before 1979), allow 150 gallons per day (gpd) per bedroom. For houses where the toilets are limited to no more than 3.5 gallons per flush and the faucets and showerheads are limited to 3 gallons per minute or less, allow 130 gpd per bedroom. For houses with water-saving toilets that use only one gallon per flush allow 90 gpd per bedroom.

The required flow rate is found by multiplying the appropriate flow by the number of bedrooms (in this case, do not count a garbage disposal as a bedroom). Knowing the rate at which water can be absorbed by the soil (the percolation rate) and the flow rate (in gallons per day), we can use Table 11-10 to calculate how many square feet of absorption field is needed. Soil with a percolation rate less than 1 minute per inch or more than 60 minutes per inch is unsuitable for a conventional system. Required Area (square feet) = Flow Rate (gallons per day) divided by the Application Rate (gallons per day per square foot). Now that the number of needed square feet of absorp-

Table 11-10 Absorption area required per bedroom, corresponding to percolation time.

Percolation rate per inch	Square feet of absorption are per bedroom
1 minute	70
2 minute	85
3 minute	100
4 minute	115
5 minute	125
10 minute	165
15 minute	190
30 minute	250
45 minute	300
60 minute	330

tion field is known, divide by the width of each trench to see how many feet of trench is required. The normal trench width is two feet.

Example: Assume a 3-bedroom house with all the latest water-saving toilets, showerheads, sinks, etc. A percolation test reveals that the percolation rate is 32 minutes per inch. How big an absorption field will be needed? The flow rate is 3 bedrooms times 90 gallons per day per bedroom, or 270 gallons per day (3 × 90 = 270). From Table 11-10, the application rate is 0.5 gallons per day, per square foot, for a percolation rate of 32 minutes per inch. The required trench area is then 270 gallons per day divided by 0.5 gallons per day per square foot (270 ÷ 0.5 =540). The required absorption area for this example would be 540 square feet. If the absorption trenches are 2 feet wide, you will need a total of 270 feet (540 ÷ 2) of absorption trench. Most health codes limit the length of any one trench (called a lateral) to no more than 60 feet. The minimum number of laterals needed for this example is calculated by dividing 270 feet by 60 feet per lateral (270 ÷ 60 = 4½), or 4.5 laterals.

Where property conditions permit, it is best to keep the laterals the same length, so in general, 5 laterals, each 60 feet long, would be specified. If there is only room on the property for laterals that are 45 feet long, simply divide 270 feet by 45 (length of lateral), which yields a total of 6 laterals (270 ÷ 45 = 6).

In addition to the area needed for the leach field, room for possible expansion should be built into the plan. Some states require contingencies for up to 50 percent expansion.

Alternative systems

The septic systems discussed so far have been of conventional system design, installed in the soil that exists on the site. Where site conditions do not lend themselves to installation of this type of system, there are alternatives. For example, if ground water or percolation rates are unsuitable, it may be possible to install a "mound" system. In a mound system, suitable soil is mounded on top of the unsuitable soil. A conventional system is then installed in the mound. There are some additional requirements for this type of design.

If there is not enough room for a conventional leach field, it may be possible to install one or more cesspools, or seepage pits. These units are usually round, require less open ground, and are deeper than a conventional leach field. Again, there are specific requirements for these systems.

Conventional, mound, and seepage pit systems all work by anaerobic bacterial action. This means the bacteria work without oxygen. Some systems are designed to be aerobic—meaning the bacteria need oxygen (air); there are also hybrid systems that use a combination of anaerobic and aerobic sections. A design professional decides if one of the nonconventional systems best suits the needs of the project.

Over the years it has become customary to calculate the required square foot of absorption area based upon the number of bedrooms in a residence. Therefore, Table 11-10 supplies the number of square feet of absorption area required per bedroom, corresponding to the time in minutes required for the water level in the hole to decrease by 6 inches.

Once the total square footage of the of the trench has been determined, by multiplying the appropriate figure from Table 11-10 by the number of bedrooms in the structure, the length of the trench(s) must be determined. Dividing the total square foot area required by the width of the intended trench(s), with the required amount of filter media below the pipe (normally 6 inches minimum), will result in the correct length of the trench(s). For example, a trench 2 feet wide with the appropriate depth of filter media would need to be 105 feet long to accommodate 210 square feet of required absorption area.

The next step is calculating the depth at which to set the tank. The tank needs to be set so as to afford a downward pitch of ¼ inch per foot of run minimum to ensure proper drainage. The depth of the disposal field tiles is another controlling factor in the placement of the tank. The outlet from the tank is generally 1 to 3 inches below the inlet, requiring the tank to be set level in its hole to avoid any rush of contents offsetting the differential between inlet and outlet.

The depth of the hole must be gaged by the practicalities of the system as well. The top of the tank should not be any more than 3 feet under the sod to facilitate servicing, but should be at least 1 foot below sod level.

Review questions

1. What is the length of pipe for the expansion bend shown in Fig. 11-11, which has a radius (R) of 24 inches, two 10 inch tangents (T), and distance in degrees (D) as shown.

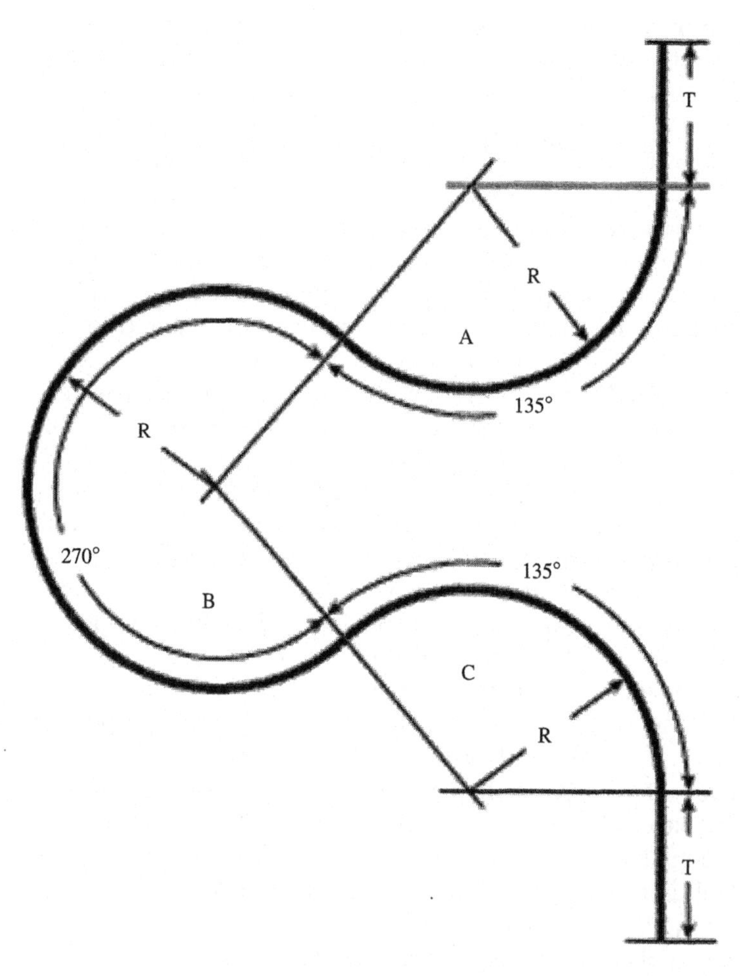

11-11

2. Approximate the number of gallons of water remaining in the horizontal tank (Fig. 11-12) by measuring the height of the liquid and using the data below.

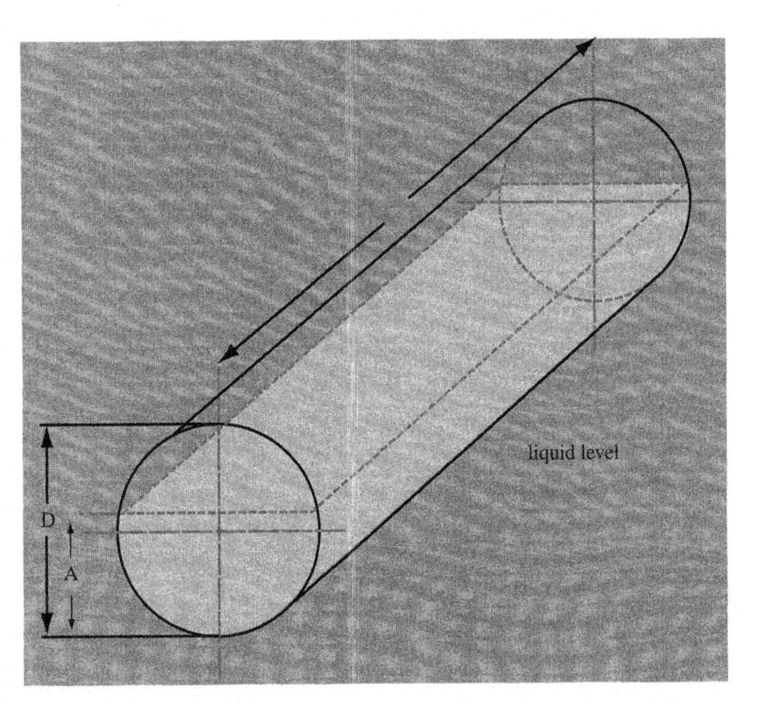

11-12

- G = gallons of water
- H = height of liquid (2 ft 9 in)
- D = diameter of tank (4 ft)
- L = length of tank (10 feet)
- K = H/D (0.69)
- P = constant for K (.7360)

3. A pipe transfers 236 gallons in 2 hours. How long will it take to fill a tank that holds 4543 gallons?
4. How much is available pressure reduced in a residential system with a difference in height of 30 feet?
5. What is the velocity of 10 gpm going through a ½″ diameter schedule 40 pipe?

12

Math for the electrician

Due to the many different applications of electricity within a structure, an abundance of mathematics is needed to plan for and properly provide for those applications.

Essential math

Without delving into the vast array of complex theory, or the complicated formulas prevalent throughout the study and application of electricity, this chapter deals with the everyday, practical uses of math by electricians.

The importance of understanding electrical terminology cannot be overemphasized. Recapping a few key words and phrases here will ensure that everyone starts with the same understanding. For example, electrical *current* flows, under pressure, through wiring. That flow would come to an absolute halt if there were no demand—that is if every single appliance, light, and other device were unplugged or otherwise disconnected from the electrical system of the building.

The amount of current (measured in *amps*) going through a wire at a given time is based on the number of electrons passing a given point each second. The pressure forcing these electrons along their route is known as *voltage* and is measured in volts.

If there were to be an increase in the voltage entering the circuit, the flow of the current would not accelerate because current travels at a constant rate—the speed of light. What would be increased would be the *power* in the lines; that power is measured in watts and is the product of multiplying the number of amps times the number of volts. The course of current flow from the source of the circuit to a plug outlet, for example, involves the electrons traveling through the *hot wire*, which normally is black. After the current has flowed through a light or appliance, the electrons seek a direct route to *ground* and travel in white neutral wires to this end. These neutral wires, also known as *system grounds*, complete every circuit in the system by returning its current to the ground.

Modern circuits have three wires, the third being a bare or green one which serves as an *equipment ground*. The role of the equipment ground wire is to ground all metal parts throughout the installation, such as conduit, armored cable, motors, and major appliances. This grounding wire protects against the danger of short circuits.

By using the formula "amps equals watts divided by volts," the amperage requirements of each electrical device on each circuit can be determined. Most residential circuits are 120 volts, and a wattage figure or amperage figure for every bulb, appliance, or other electrical device should be calculable. If a wattage figure is present on the device or on a label on the device, dividing that wattage figure by 120 will yield the amperage for the device.

Adding the amperage figures for each circuit and comparing that total with the amperage capacity appearing on the appropriate breaker or fuse will indicate whether a circuit would face an overload if every device tied into it called for current at the same time.

Ohm's Law

Written as an equation, Ohm's law states

$$I = \frac{E}{R}$$

where I represents the flow of current expressed in amperes, E stands for the amount of voltage (in volts), and R represents the amount of resistance present in the circuit.

In English, this equation states that the current flowing through an electrical circuit is directly proportional to the applied voltage while being inversely proportional to the total resistance within the circuit.

Calculating voltage, amperage, & resistance

By substituting the values of any two know factors of Ohm's law into the equation and preforming the calculations, determining the quantity of the one remaining factor is easy. For example, if a circuit is being fed 12 volts (V) and has a resistance of 6 ohms (Ω) the amount of current flow (in Amps) within the circuit would be calculated as follows:

$$I = \frac{E}{R}$$

$$I = \frac{12}{6}$$

$$I = 2 \text{ amps}$$

To illustrate the direct relationship between the amount of voltage applied to the circuit and the current flowing through the circuit consider the equation above substituting twice the applied voltage.

$$I = \frac{E}{R}$$

$$I = \frac{24}{6}$$

$$I = 4 \text{ amps}$$

The amount of current flowing through the circuit changes in the same direction and by the same factor as the quantity of applied voltage. The amount of voltage increased by a factor of 2 and the amount of current in the circuit increased by a factor of 2.

To illustrate the inverse relationship existing between the amount of current in the circuit and the resistance within the circuit, consider the same equation as given previously, but with the resistance decreased to 3 Ω.

$$I = \frac{E}{R}$$

$$I = \frac{12}{3}$$

$$I = 4 \text{ amps}$$

Notice that, with the amount of resistance decreased by a factor of 2 ($6 \div 2 = 3$), the amount of current flow in the circuit increased by a factor of 2. The change in the quantity of current flow moves in opposition to the change in the quantity of resistance in the circuit; therefore, it is an inverse relationship.

Ohm's law is an algebraic equation, a statement that stands on its own presenting a truth about the relationship between the three factors of the equation. As it has been stated, the equation is designed to "solve" for the missing value of I (current). If the amount of voltage (E) were the missing factor to be determined, the equation would have to be rearranged to indicate it. Rearranging the equation is accomplished through a method called *transposition.* In common-sense terms, when the equation was set up to calculate the value of I, I was isolated on one side of the equal sign in the equation. If E is the factor to be solved for, then E needs to be isolated on one side of the equation instead of I. To accomplish this start with the original equation

$$I = \frac{E}{R}$$

To isolate E, something must be done to remove R from the same side as E is on. Because the original equation calls for E to be divided by R, if E were to be multiplied by R, then R would be removed from that side of the equation, leaving E isolated on that side of the equals sign. As with all algebraic equations, whatever is done to one side of the equation must be done equally to the other side of the equation. Therefore, I must also be multiplied by R, which results in that side of the equation becoming IR (or $I \times R$) instead of just I. The new (transposed) equation now reads:

$$IR = E$$

which in English translates to the statement "current times resistance is equal to voltage." Plugging in the values for the two know factors and performing the appropriate calculations, the equation becomes

$$4 \text{ amps} \times 3\ \Omega = E$$

$$12 = E$$

$$E = 12 \text{ volts}$$

If the amount of resistance (R) were to be the value to be determined, the equation again needs to be transposed to reflect this. Following the same procedure as previously described to transpose the equation to the point where it reads

$$IR = E$$

the R must now be isolated on one side of the equation. Because the equation now reads I times R (or R multiplied by I), the way to remove I from that side of the equation is to divide R by I, which leaves just R. Again, both sides of the equation must be treated equally, so E is also divided by I which results in the formula now reading

$$R = \frac{E}{I}$$

Plugging in the values for the known elements of the equation and performing the calculations will yield the following:

$$R = \frac{E}{I}$$

$$R = \frac{12 \text{ volts}}{2 \text{ amps}}$$

$$R = \frac{12}{2}$$

$$R = 6\ \Omega$$

Wire types & sizes

Electrical wires used for general purpose residential applications are identified by numbers ranging from 000 to 14. An inverse relationship exists between the size of the wire and its identifying number; number 000 wire is considerably larger than number 14 wire. Larger wires do exist and are used in heavy industrial applications to carry extremely high voltages while wires which are smaller than number 14 wire are utilized in applications such as low voltage control circuits.

The three types of wiring most commonly used in construction are BX, Romex, and EMT. BX is a trade name by which this flexible armored cable has become synonymous; likewise, Romex is also a trade name widely used to refer to this type of plastic-covered cable, while EMT stipulates a thin-walled conduit. In addition to the three types of wire mentioned, some variations are useful and permitted by code.

- *Non-metallic cable.* Most local codes allow the use of non-metallic sheathed cable inside walls, floors, and other places where it cannot be damaged. Information listed on the plastic covering of the cable details the contents of the cable. Type NM, also called Romes (another trade name), is designed for use in dry locations only, while type NMC is designed to be run in dry locations or damp locations. A third type of non-metallic cable is UF (underground feeder). Because UF cable is specifically designed for use in wet locations, this cable can be completely buried.
- *BX cable.* The flexible armored cable, known as BX has a spiral-wrapped steel cover. It can be used in dry or damp places and, in some instances, for short exposed runs. In many localities, armored cable and flexible conduit are orphans. They might be found in an older home, but are usually not permitted to be used for new work. A few other communities insist that only armored cable is run, not non-metallic cable. Check the local codes in each area for the exact regulations governing the use of this material.

- *Flexible steel conduit.* Flexible steel conduit or "Greenfield" looks like armored cable without the wires. The conduit can be cut to length, have the wires threaded through it, and then can be installed in completed pieces.
- *Thinwall and rigid conduit.* With thinwall and rigid conduit, wires are fished through from their origin to their end point after the conduit has been installed. Most codes require the use of these types of conduits for exposed runs.
- *Aluminum or copper-clad wiring.* In some localities, running aluminum or copper-clad aluminum wiring is permissible. Be warned, however, that aluminum expands and contracts considerable more than copper. This difference in the rate of expansion between the two materials can result in a loosening at the terminals. Consequently, only devices marked CO/ALR or CU/AL should be used in conjunction with aluminum or copper-clad aluminum wiring.

Electrical wiring is coded to reflect the coating (composition) of the wire and its purpose. The following codes and their meanings are presented as an example of some of the most commonly used wire types.

R	Rubber
RW	Moisture-resistant rubber
TW	Moisture-resistant thermoplastic
RH	Heat-resistant rubber
RU	Latex rubber
T	Thermoplastic
RHW	Moisture- and heat-resistant rubber

Table 12-1 presents the maximum amperage carrying capacity of a variety of different wire sizes, categorized by their code numbers.

The actual size of a wire is usually described in terms of mils. A mil is equal to one thousandths of an inch (1/1000). Since the shape of a wire is normally round, the term *mil* is converted for use with wires and is called a *circular mil* in this case. A circular mil refers to the circular cross-section of a wire with a diameter measuring 1/1000 of an inch. Converting the area of a circle into circular mils (CM) is done by taking the di-

Table 12-1 Maximum amperage load per wire size.

Size of wire (wire number)	Max. amps for wires coded R, RW, RU, T and TW	Max. amps for wires coded RH, RHW
000	165	200
00	145	175
0	125	150
1	110	130
2	95	115
3	80	100
4	70	85
6	55	65
8	40	45
10	30	30
12	20	20
14	15	15

ameter of the circle (in mils) and multiplying that figure by itself (squaring the figure). The mathematical equation for this would appear as:

$$A = D^2$$

where A is the area to be determined in circular mils and D is equal to the diameter of the circle measured in mils. If the diameter of the circle is 300 mils, the calculations would be as follows:

$$A = D^2$$

$$A = 300^2 \text{ (or } 300 \times 300)$$

$$A = 90{,}000 \text{ circular mils}$$

Fiber optics

Since its invention in the early 1970s, the use and demand of optical fiber has grown tremendously. The uses of optical fiber today are quite numerous. The most common are telecommunications, medicine, military, automotive, and industrial. Telecommunications applications are widespread, ranging from global networks to local telephone exchanges to subscribers' home telephone service and desktop computers. The transmission of voice, data, audio and video over distances of less than a meter to hundreds of miles are common applications using one of a few standard fiber designs in one of several cable designs.

Companies such as AT&T, MCI, and U.S. Sprint use optical fiber cable to carry plain old telephone service (POTS) across their nationwide networks. Local providers use fiber to carry telephone service between central office switches at local levels, and sometimes extend as far as neighborhood nodes or into individual homes. The use of optical fiber for data transmission is seen extensively in firms such as IBM, Rockwell, Honeywell, as well as many banks, universities, Wall Street firms. These firms have a need for secure, reliable systems to transfer computer and monetary information between buildings to desktop terminals around the world. The security inherent in optical fiber systems provides a major benefit.

The cable television industry has found fiber optics useful in delivering video services. The high information-carrying capacity (bandwidth) of fiber makes it the perfect choice for transmitting signals to and from subscribers.

Basic principles of operation

In simple terms, fiber optics is a medium for carrying information from one point to another in the form of light. Unlike copper-based transmission, fiber optics is not electrical in nature. The basic fiber optic system consists of three components:

1. A transmitting device, which generates the light signal
2. An optical fiber cable, which carries the light
3. A receiver, which accepts the light signal transmitted.

The fiber itself is passive and does not contain any active, generative properties.

As seen in Figure 12-1, the optical fiber itself consists of three components:

1. core
2. cladding
3. coating

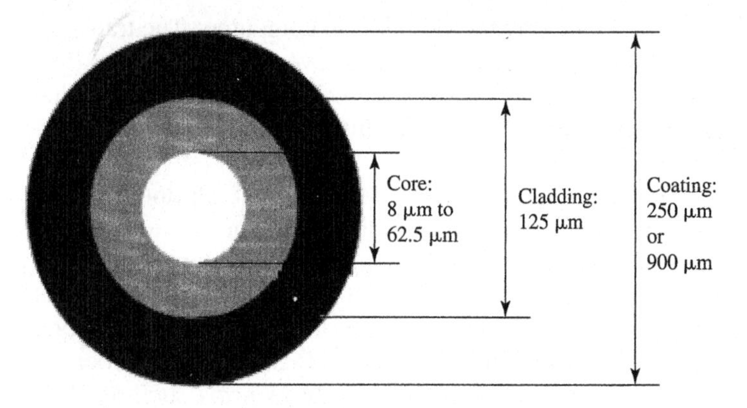

12-1 *Cross section of typical optical fiber.*

- The core is the central region of an optical fiber through which light is transmitted. In general, the telecommunications industry uses sizes from 8.3 micrometers (µm) to 62.5 micrometers. The standard telecommunications core sizes in use today are 8.3 µm (single-mode), 50 µm (multimode), and 62.5 µm (multimode).
- The diameter of the cladding surrounding each of these cores is 125 µm. Core sizes of 85 µm and 100 µm have been used in early applications, but are not typically used today. To put these sizes into perspective, compare them to a human hair, this is approximately 70 µm or 0.003 inch. The core and cladding are manufactured together as a single piece of silica glass with slightly different compositions, and cannot be separated from one another. The glass does not have a hole in the core, but is completely solid throughout.
- The third section of an optical fiber is the outer protective coating. This coating is typically an ultraviolet (UV) light-cured acrylate applied during the manufacturing process to provide physical and environmental protection for the fiber. During the installation process, this coating is stripped away from the cladding to allow proper termination to an optical transmission system. The coating size can vary, but the standard sizes are 250 µm or 900 µm. The 250 µm coating

takes less space in larger outdoor cables. The 900 μm coating is larger and more suitable for smaller indoor cables.

Once light enters an optical fiber, it travels in a stable state called a mode. There can be from one to hundreds of modes depending on the type of fiber. Each mode carries a portion of the light from the input signal. The number of modes in a fiber is a function of the relationship between core diameter, numerical aperture, and wavelength.

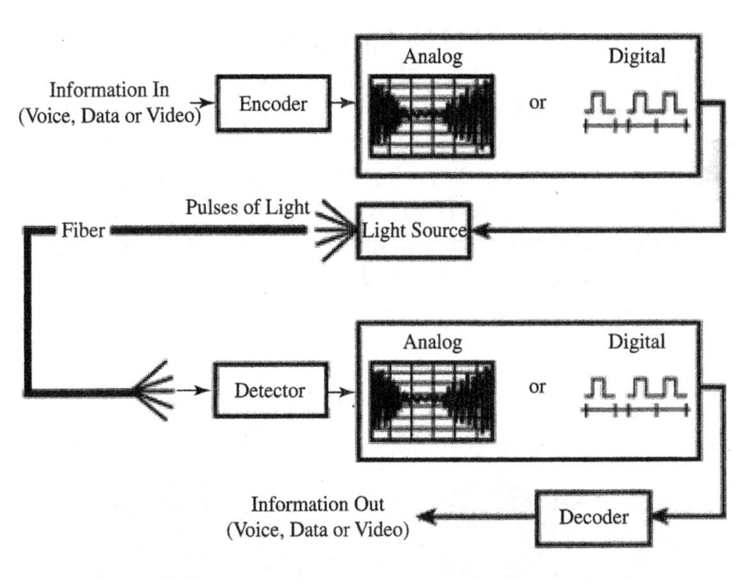

12-2 *Sequence of events within a fiber-optic system.*

Information transmission sequence

Figure 12-2 demonstrates the sequence of events within a fiber optic system. First, information (voice, data, or video) must be encoded into electrical signals. At the light source, these electrical signals must next be converted into light signals. Contrary to popular belief that fiber can transmit only digital signals due to the on/off binary characteristic of the light source, it is important to note that fiber has the capability to carry either analog or digital signals. The intensity of the light and the frequency at which the intensity changes can be used for AM and FM analog transmission.

Once the signals are in the form of light, they travel down the fiber until they reach a detector, which changes the information from light signals back into electrical impulses. The area from the light source to the detector constitutes the passive transmission subsystem. Finally, the electrical signals are decoded back into the form of voice, data, or video information.

Types of Fiber

Every telecommunications fiber falls into one of two categories: single-mode or multimode. It is impossible to distinguish between single-mode and multimode fiber with the naked eye. There is no difference in outward appearance, only in core size (see Figure 12-3). Both fiber types act as a transmission medium for light, but they operate in different ways, have different characteristics, and serve different applications.

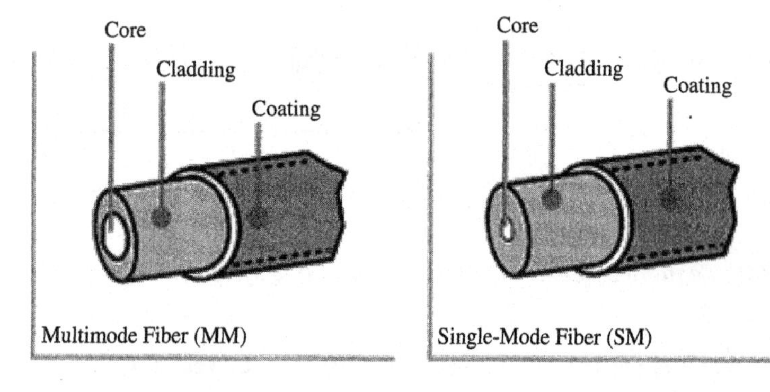

12-3 *Single-mode and multimode fiber.*

Single-mode fiber

Single-mode (SM) fiber allows for only one pathway, or mode, of light to travel within the fiber. The core size is typically 8.3 μm. Single-mode fibers are used in applications where low signal loss and high data rates are required, such as on long spans where repeater/amplifier spacing needs to be maximized.

Multimode fiber

Multimode (MM) fiber allows more than one mode of light to travel within the fiber at the same time. Common MM core sizes are 50 μm and 62.5 μm. Multimode fiber is better suited for shorter distance applications. Where costly electronics are heavily concentrated, the primary cost of the system does not lie with the cable. In such a case, MM fiber is more economical because it can be used with inexpensive connectors and LED transmitters, making the total system cost lower. This makes MM fiber the ideal choice for short distance, lower bandwidth applications.

Applied principles of operation

As with any type of transmission system, certain parameters affect the system's operation. Light that is seen by the unaided human eye falls within the visible light spectrum. In the visible spectrum, wavelengths are identified using the color of light typical to each wavelength.

Optical fiber transmission uses wavelengths which are above the visible light spectrum, and thus undetectable to the unaided eye. Typically, 850 nanometers (nm), 1310 nm, and 1550 nm are used as optical transmission wavelengths.

Both lasers and LEDs (light-emitting diodes) are used to transmit light through optical fiber. Lasers are usually used for 1310 or 1550 nanometer, single-mode applications. LEDs perform well in 850 or 1300 nanometer multimode applications.

- **Safety note: Never look into the end of a fiber that may have a laser coupled to it. Laser light is invisible and can damage the eyes. Viewing it directly does not cause pain. The iris of the eye will not close involuntarily as when viewing a bright light; consequently, serious damage to the retina of the eye is possible. Accidental exposure to laser light should be followed by an eye examination as quickly as possible.**

Operating windows

Operating windows are ranges of wavelengths at which the fiber will operate at peak efficiency. Each window is centered on the typical operational wavelength.

The windows and their corresponding operating wavelengths shown in Table 12-2 were chosen because they best match the transmission properties of available light sources with the transmission qualities of optical fiber.

Table 12-2 Matching windows and wavelengths.

Window	Operating wavelength
800–900 nm	850 nm
1250–1350 nm	1310 nm
1500–1600 nm	1550 nm

Frequency

The frequency of a system is the speed of modulation of the digital or analog output of the light source; in other words, the number of pulses per second emitted from the light source. Frequency is measured in units of hertz (Hz), where 1 hertz is equal to 1 pulse or cycle per second (Figure 12-4). A more practical measurement for optical communications is megahertz (MHz) or millions of pulses per second.

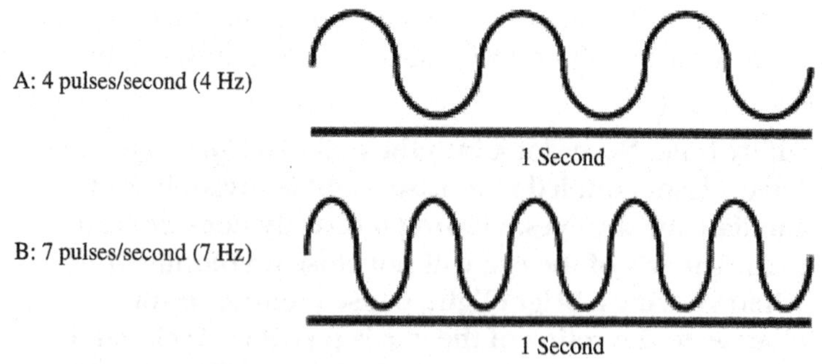

12-4 *1 hertz is equal to 1 pulse or cycle per second.*

Index of refraction

The index of refraction (IOR) is a way of measuring the speed of light within a material. The actual speed of light, given as 186,000 miles per second (300,000 kilometers per second) is calculated within a vacuum, such as outer space. Light traveling through any medium (water, glass, plastic etc.) will travel slower than in a vacuum. Index of Refraction is calculated by dividing the speed of light in a vacuum by the speed of light in some other specified medium.

The Index of Refraction of a vacuum by definition has a value of 1. For quick reference, Table 12-3 lists the typical values of some common media. The larger the index of refraction, the more slowly light travels in that medium. Notice that the typical value for the cladding of an optical fiber is 1.46. The value for the core is 1.48. To calculate the index of refraction within the fiber core the formula would be:

$$IO = (\text{speed of light in a vacuum}) \div (\text{speed of light within core})$$

$$IO = (186{,}000 \text{ miles per second}) \div 1.48$$

$$IO = 127{,}397 \text{ miles per second}$$

Fighting its way through the core has slowed the speed of the light wave by approximately 58,603 miles per second. While a change of this speed is not noticeable to the human eye, the data detector will be able to sense a change. If this change is carried over too much of a distance the reader will not be able to properly determine the message.

Table 12-3 Index of refraction.

Medium	Typical index of refraction (infrared light)	Speed
Vacuum	1.0000	Faster
Air	1.0003	↑
Water	1.33	
Cladding	1.46	↓
Core	1.48	Slower

Total internal reflection

When a light ray traveling in one material hits a different material and reflects back into the original material without any loss of light, total internal reflection occurs. Since the core and cladding are constructed from different compositions of glass, theoretically, light entering the core is confined to the boundaries of the core because it reflects back whenever it hits the cladding. For total internal reflection to occur, the index of refraction of the core must be higher than that of the cladding (as in the previous example).

Acceptance cone

To ensure that the light pulses (signals) travel correctly through the core, they must enter the core through what is known as an acceptance cone (see Figure 12-5). There is a angle from the fiber axis at which light may enter the fiber which maximizes its travel in the core of the fiber. The sine (see Table 12-4) of this maximum angle is the numerical aperture (NA) of the fiber. The size of the acceptance cone is a function of the refractive index difference between the core and the cladding. Fiber with a larger NA requires less precision to splice and work with than fiber with a smaller NA.

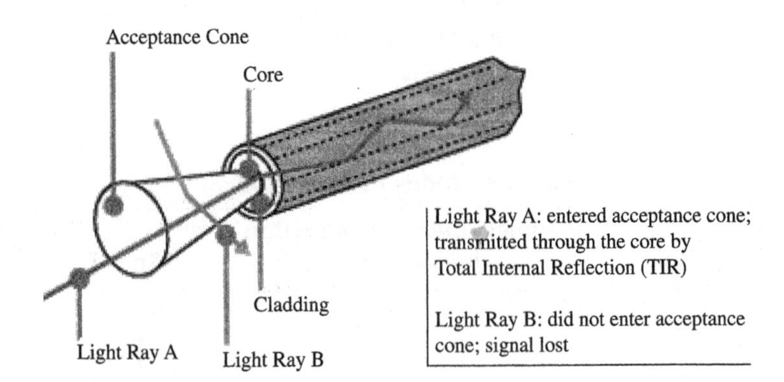

12-5 *Example of an acceptance cone.*

Table 12-4 Table of sines.

Radian	Degree	Sine
.0000	00	.0000
.0175	01	.0175
0.349	02	.0349
.0524	03	.0523
.0698	04	.0698
.0873	05	.0872
.1047	06	.1045
.1222	07	.1219
.1396	08	.1392
.1571	09	.1564
.1745	10	.1736
.1920	11	.1908
.2094	12	.2079
.2269	13	.2250
.2443	14	.2419
.2618	15	.2588
.2793	16	.2756
.2967	17	.2924
.3142	18	.3090
.3316	19	.3256
.3491	20	.3420
.3665	21	.3584
.3840	22	.3746
.4014	23	.3907
.4189	24	.4067
.4363	25	.4226
.4538	26	.4384
.4712	27	.4540
.4887	28	.4695
.5061	29	.4848
.5236	30	.5000
.5411	31	.5150
.5585	32	.5299
.5760	33	.5446
.5934	34	.5592
.6109	35	.5736
.6283	36	.5878
.6458	37	.6018
.6632	38	.6157
.6807	39	.6293
.6981	40	.6428
.7156	41	.6561
.7330	42	.6691
.7505	43	.6820
.7679	44	.6947
.7854	45	.7071

Attenuation

Attenuation is the loss of optical power as light travels down a fiber. Measured in decibels (dB/km) over a set distance, a fiber with a lower attenuation will allow more power to reach its receiver than a fiber with higher attenuation. While low-loss optical systems are always desirable, it is possible to lose a large portion of the initial signal power without significant problems. A loss of 50 percent of initial power is equal to a 3.0 dB loss. Any time fibers are joined together there will be some loss. Losses for fusion splicing and for mechanical splicing are typically 0.2 dB or less. Attenuation can be caused by several factors, but is generally placed in one of two categories: intrinsic or extrinsic.

- **Intrinsic attenuation** occurs due to something inside or inherent to the fiber. It is caused by impurities in the fiber during the manufacturing process. As precise as manufacturing is, there is no way to eliminate all impurities, though technological advances have caused attenuation to decrease dramatically since the first optical fiber in 1970.

When a light signal hits an impurity in the fiber, one of two things will occur: it will scatter or it will be absorbed.

- **Scattering:** Rayleigh scattering accounts for the majority (about 96 percent) of attenuation in optical fiber. Light travels in the core and interacts with the atoms in the fiber. The light waves elastically collide with the atoms, and light is consequently scattered. Rayleigh scattering is the result of these elastic collisions between the light wave and the atoms in the fiber. If the scattered light maintains an angle that supports forward travel within the core, no attenuation occurs. If the light is scattered at an angle that does not support continued forward travel, the light is diverted out of the core and attenuation occurs as seen in Figure 12-6.

Some scattered light is reflected back toward the light source. This property is used to test fibers using an Optical Time Domain Reflectometer (OTDR). This same principle applies to analyzing loss associated with localized events in the fiber, such as splices.

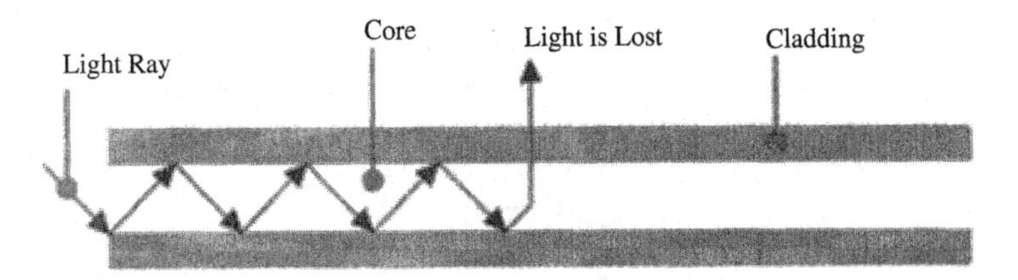

12-6 *Light scattering within the fiber.*

- **Absorption:** The second type of intrinsic attenuation in fiber is absorption. Absorption accounts for 3-5 percent of fiber attenuation. This phenomenon causes a light signal to be absorbed by natural impurities in the fiber, and converted to vibrations or some other form of energy (Figure 12-7). Unlike scattering, absorption can be limited by controlling the amount of impurities during the manufacturing process.

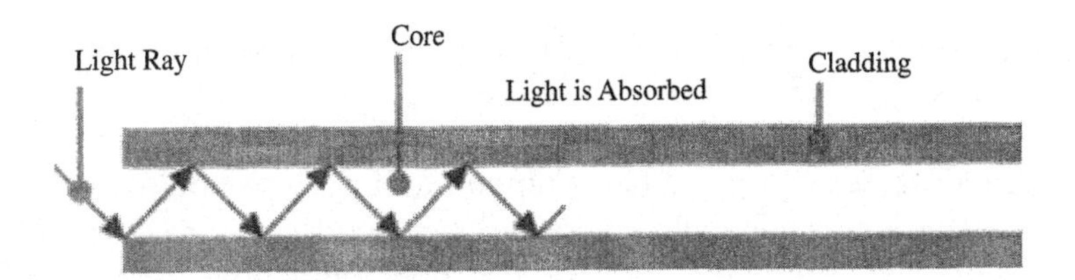

12-7 *Light absorption within a fiber.*

- **Extrinsic attenuation** can be caused by two external mechanisms: macrobending or microbending. Both cause a reduction of optical power.

Macrobending

If a bend is imposed on an optical fiber, strain is placed on the fiber along the region that is bent (Figure 12-8). The bending strain will affect the refractive index and the critical angle of

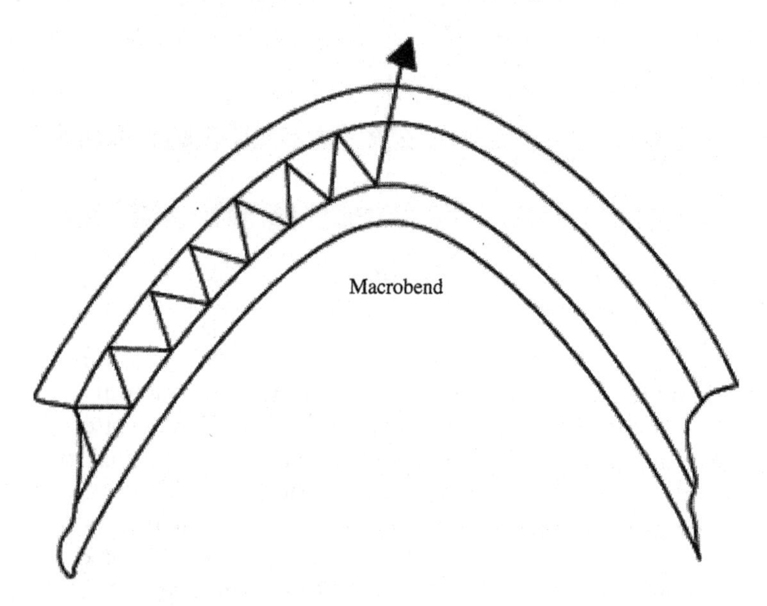

12-8 *Typical example of a macrobend.*

the light ray in that specific area. Consequently, light traveling in the core can refract out, and loss occurs.

A macrobend is a large-scale bend (Figure 12-8) that is visible; for example, a fiber wrapped around a person's finger. This loss is generally reversible once bends are corrected. To prevent macrobends, all optical fiber (and optical fiber cable) has a minimum bend radius specification that should not be exceeded. This is a restriction on how much bend a fiber can withstand before experiencing problems in optical performance or mechanical reliability. The rule of thumb for minimum bend radius is 1-1/2 inches for bare, single-mode fiber; 10 times the cable's outside diameter (O.D.) for non-armored cable; and 15 times the cable's O.D. for armored cable.

Microbending

The second extrinsic cause of attenuation is a microbend. This is a small-scale distortion, generally indicative of pressure on the fiber. (See Figure 12-9). Microbending may be related to temperature, tensile stress, or crushing force. Like macrobending, microbending will cause a reduction of optical power in the fiber. Microbending is usually a localized event, and the bend may not be clearly visible upon inspection.

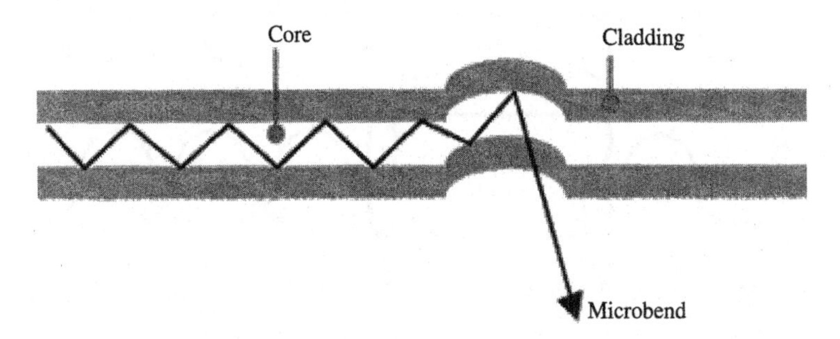

12-9 *Microbending will cause a reduction of optical power.*

Dispersion

Dispersion is the "spreading" of a light pulse as it travels down a fiber. As the pulses spread, or broaden, they tend to overlap, and are no longer distinguishable by the receiver as 0s and 1s. High data rates which launch light pulses close together spread too much resulting in a high dispersion rate. This high dispersion rate results in errors and loss of information.

Chromatic dispersion is a result of the range of wavelengths in the light source. Laser and LED generated light consists of a range of wavelengths. Each of these wavelengths travels at a slightly different speed. Over distance, the varying wavelength speeds cause the light pulse to spread.

Modal dispersion is significant in multimode applications, where the various modes of light traveling down the fiber arrive at the receiver at different times, causing a spreading effect. The spreading of these light pulses causes them to merge. At a certain distance and frequency, the pulses become unreadable by the receiver (Figure 12-10). The multiple pathways of a multimode fiber cause this overlap to be much greater than for single-mode fiber. These different paths have different lengths, which cause each mode of light to arrive at a different time.

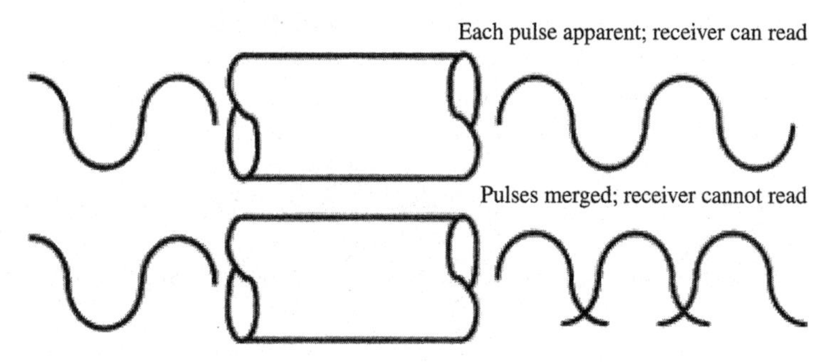

12-10 *Typical example of modal dispersion.*

In all multimode optical fiber, there is a direct correlation between the value of measured bandwidth and the dispersion that occurs when light pulses propagate along the fiber. As the modal dispersion increases, the bandwidth (AKA. modal bandwidth) decreases. This decrease in bandwidth will limit the distance a pulse will be able to travel along the core and still be recognized at the receiver as the original signal.

Bandwidth

Bandwidth is the measure of the information carrying capacity of the medium in question. For copper cables currently in use in LAN networks, bandwidth is very much dependent upon the frequency at which the signal is transmitted. As the frequency of the signal increases, less bandwidth (and higher attenuation) will be evident in the systems.

Other factors will affect the bandwidth of optical fibers. For example, the basic mechanism affecting the bandwidth of optical fibers is the dispersion, or spreading, of the light pulse as it travels down the fiber core. Bandwidth is length dependent for optical fibers. The longer the path taken over the absolute distance, the less likelihood that the optical signals will arrive at the detector within the allotted time slot and in the correct order to be properly interpreted. Dispersion limits how fast, or how much, information can be sent over an optical fiber.

System bandwidth is measured in megahertz (MHz) at a length of one km. In general, when a system's bandwidth is 200 MHz·km, it means that 200 million pulses of light per second will travel down 1 km (1000 meters) of fiber, and each pulse will be distinguishable by the receiver.

How distance is impacted by bandwidth can be calculated using the following formula:

L = (BW) ÷ (DR × BN)

L = Fiber length in meters

BW = Bandwidth of fiber

DR = System data rate

BN = *Bandwidth normalized to the bit rate

* BN will be .71 for multimode fiber

Example:

$$L = (160 \text{ MHz}) \div (1 \text{ GHz} \times .71)$$

$$L = 225.35 \text{ meters}$$

Consequently, if bandwidth is increased to 200 MHz the equation now becomes:

$$L = (200\text{MHz}) \div (1 \text{ GHz} \times .71)$$

$$L = 281.69 \text{ meters}$$

On the other hand, if the data rate is increased length is decreased as shown in the following example.

$$L = (200 \text{ MHz}) \div (2 \text{ GHz} \times .71)$$

$$L = 140.84 \text{ meters}$$

Any increase in the fiber's bandwidth will correlate to a direct increase in the length of fiber that will support a given data rate. Any increase in data rate not supported by a proportional increase in bandwidth will result in a decrease in fiber length. **Note: Small variations will occur when using this formula due to the fact that the BN is normalized using an average system performance and not any one specific system.**

Fiber v metal

Optical fiber systems have many advantages over metallic-based communication systems. These advantages include:

- Long distance signal transmission
  The low attenuation and superior signal integrity found in optical systems allow much longer intervals of signal transmission than metallic-based systems. While single-line, voice-grade copper systems longer than a couple of kilometers (1.2 miles) require in-line signal repeaters for satisfactory performance, it is not unusual for optical systems to go over 100 kilometers (km), or about 62 miles, with no active or passive processing. Emerging technologies promise even greater distances in the future.
- Large bandwidth, light weight, and small diameter
  While today's applications require an ever-increasing amount of bandwidth, it is important to consider the space constraints of many end-users. It is commonplace to install new cabling within existing duct systems. The relatively small diameter and light weight of optical cables makes such installations easy and practical, and saves valuable conduit space in these environments.
- Long lengths
  Long, continuous lengths also provide advantages for installers and end-users. Small diameters make it practical to manufacture and install much longer lengths than for metallic cables: twelve-kilometer (12 km) continuous optical cable lengths are common. Manufacturers produce continuous single-mode cable lengths up to 12 km, with a reel size of 96-inches being the primary limiting factor. Multimode cable lengths can be 4 km or more, although most standards require a maximum length of 2 km or less. Multimode cable lengths are based on industry demand.
- Easy installation and upgrades
  Long lengths make optical cable installation much easier and less expensive. With some modifications due to the small size and limited pull tension and bend radius of optical cables the same equipment used to install copper and coaxial cables can be used to install optical fiber cables. Depending on the duct system's condition, layout, and the installation technique, optical cables can typically be

installed in spans of 6000 meters or more. The longer cables can be coiled at an intermediate point and pulled farther into the duct system as necessary.
System designers typically plan optical systems that will meet growth needs for a 15- to 20-year span. Although sometimes difficult to predict, future expansion is often planned for by installing spare fibers at the time of installation. Installation of spare fibers today is more economical than installing additional cables later.

- Non-conductivity
Another advantage of optical fibers is their dielectric nature. Since optical fiber has no metallic components, it can be installed in areas where electromagnetic interference (EMI), including radio frequency interference (RFI) causes problems for copper installations. Areas with high EMI include utility lines, power-carrying lines, and railroad tracks. All-dielectric cables are also ideal for areas of high lightning-strike incidence.
- Security
Unlike metallic-based systems, the dielectric nature of optical fiber makes it impossible to remotely detect the signal being transmitted within the cable. The only way to do so is by actually accessing the optical fiber itself. Accessing the fiber requires intervention that is easily detectable by security surveillance. These circumstances make fiber extremely attractive to governmental bodies, banks, and others with major security concerns.
- Designed for future applications needs
Fiber optics is affordable today, as electronics prices fall and optical cable pricing remains low. In many cases, fiber solutions are less costly than copper. As bandwidth demands increase rapidly with technological advances, fiber will continue to play a vital role in the long-term success of telecommunications.

Current and future standards

The fiber industry has established standards for fiber performance to ensure that customers consistently receive the quality they expect (Table 12-5). For a standard multimode fiber with a core/clad of 62.5/125 microns and a numerical aperture of 0.275, the standard FDDI bandwidth is 160/500 MHz·km (at

Table 12-5 Applications standards summary.

	*62.5 μm**		*50 μm***	
	850nm	**1300nm**	**850nm**	**1300nm**
ATM @ 622 Mb/s	300m	500m	300m	500m
Gigabit Ethernet	220m	550m	500m	550m
Fibre Channel @ 1.062 Gb/s	175m	N/A	500m	N/A
ATM @ 2.5 Gb/s	100m	300m	300m	300m

* 160/500 MHz-Km bandwidth
** 400/400 MHz-Km bandwidth

850/1300 nm). This is the most common fiber used for Ethernet and Fast Ethernet systems in North America.

Even greater transmission speed is avail-able from the newest industry protocol for data communications, called Gigabit Ethernet. This technology utilizes a laser as the light source, which, offers a narrower spectral width, resulting in less chromatic dispersion.

New fibers with enhanced bandwidth are now being introduced to take advantage of this new protocol. For example, a fully standards-compliant 62.5/125-graded index multimode fiber optimized for use in Gigabit Ethernet data communications systems has already been developed. This new fiber supports link lengths of 300 meters at 850 nm and 550 meters at 1300 nm in Gigabit Ethernet systems.

Summary

Optical fiber has become the transmission medium of choice for the high-speed data transmission and networking applications that are changing the way businesses and organizations share information. Multimode fiber's reliability and ease of handling and connectorization have made it a popular choice for local-area and wide-area networks.

Next-generation systems using Gigabit Ethernet protocols may require an enhanced-bandwidth, multimode, optical fiber for longer transmission lengths. Since this is an area of rapid technology development, network designers installing Gigabit

Ethernet today, or who will upgrade to Gigabit Ethernet in the future, should consider an enhanced-bandwidth fiber to protect their cable investments in the future.

Calculating wire size & requirements

Wires are the conductors along which the current is carried to the load (the final use to which the current will be put). In most cases the wire encountered by an electrician will be made of copper; therefore, copper would be the actual conductor. The current carrying capacity of a conductor is influenced by three major factors: the amperage, the wire size, and the wire temperature. As a rule of thumb for selecting wiring that will function at safe temperatures, the wire must have a carrying capacity greater than that of the rated amperage of the connected load.

Resistivity A material's capacity to resist the flow of an electric current through it is called resistivity. The resistivity of a particular material is measured as the area of a cross-section of the material one foot in length. As applied to wire, the cross-section is measured in circular mils, a piece of material one foot long with a cross sectional area of 1 circular mil is known as a circular mil foot (or mil foot).

When resistivity is applied to a mil foot of one specific type of material, it is known as that material's *specific resistance*. For example, a mil foot of copper has a resistivity of 10.4 Ω or, said more precisely, copper has a specific resistance of 10.4 ohms. The formula for determining the resistivity of a conductor (such as wire) is shown below:

$$R = \frac{KL}{A}$$

where R equals the material's resistivity expressed in ohms, K represents the specific resistance of the material, and L is the length of the conductor while A equals the cross-sectional area of the conductor. In practice, the electrician may have to calculate the resistance of a copper wire that has a diameter of 0.250 inches and is 1500 feet long; the necessary computations are as follows:

$$R = \frac{KL \div A = (K \times L)}{A}$$

To begin plugging numbers into the equation, the diameter of the wire must first be converted into terms of area, measured in circular mils. As stated previously the conversion is accomplished by squaring the diameter ($A = D^2$). The equation would now read

$$R = \frac{(K \times L)}{(250)^2}$$

The specific resistance of copper, as stated earlier, is 10.4 and the length of the wire is to be 1500 feet; the equation now becomes

$$R = \frac{10.4 \times 1500}{(250)^2}$$

Performing the calculations, the equation would read as follows:

$$R = \frac{15{,}600}{62{,}500}$$

$$R = 0.250 \text{ ohms}$$

Estimating wire materials & costs

As with any algebraic equation, the value of any of the elements in an equation can be determined by the equation, as long as the values of the remaining elements are known. For example, in the previous equation, if the length of the wire were the value to be determined instead of the resistance, the equation would be transposed using the same techniques as were demonstrated with Ohm's law to isolate L (length) on one side of the equation. The resulting equation would appear as:

$$L = \frac{RA}{K}$$

By plugging in the known values for R, A, and K, the equation is ready for calculation and would appear as follows:

$$L = \frac{0.250 \times (250)^2}{10.4}$$

$$L = 15{,}625 \div 10.4$$

$$L = @\ 1500 \text{ feet}$$

Table 12-6 presents the specific resistance for each of several popular materials used as conductors.

The resistance of a wire will be directly affected by the size of the wire. An inverse relationship exists between the cross sectional size of a wire (the diameter measured as circular mils, which equals the diameter measured in mils, squared) and the resistance of that wire. This relationship can be demonstrated by the fact that a wire with twice as much cross-sectional area as another wire will have half as much resistance as the smaller wire.

Incorporating values from Ohm's law into the formula used to calculate resistance will provide the means to determine the size of wire required for the circumstances. By substituting the values of R (resistance), E (voltage) and I (current) from Ohm's law, the formula for determining resistance becomes the following equation:

$$d^2 = L \times \frac{K}{\left(\frac{E}{I}\right)} \text{ or}$$

$$d^2 = L \times K \times \frac{I}{E}$$

where d^2 is the cross-sectional area of the wire (the size or gauge wire to be used), L is the length of the wire, K is the specific resistance of the material being used as a conductor, I rep-

Table 12-6 Specific resistance of different conductive materials.

Type of conductor	Specific resistance
Copper	10.4 ohms
German silver	128.3 ohms
Silver	9.8 ohms
Aluminum	17.2 ohms
Iron	63.4 ohms

resents the amount of current flowing through the circuit, and E is the amount of voltage being fed into the circuit.

Most of the time, electricians will not need to determine, by formula, the size wire needed for roughing in a building. Various state, Federal, and local government bodies, in conjunction with trade associations representing the industry, have established codes and guidelines that govern all aspects of wire size and circuit loading for all types of construction.

The use of these formulas and the knowledge they impart comes into play with regard to the various branch circuits within a structure. Table 12-7 provides the various wire sizes and types that correspond to specific amperage needs on feeder and branch circuits.

The length of wire through which a current runs will not affect the intensity of the current. Because of electricity's properties and the fact that it travels at the speed of light, a 500-foot run of wire will not have any more diminishing effect on the power at the outlet than if the outlet were one foot away from

Table 12-7 Wire sizes corresponding to amperage requirements.

	Continuous use		*Non-continuous use*	
Amperes	**Copper wire size**	**Aluminum wire size**	**Copper wire size**	**Aluminum wire size**
15	14	12	14	12
20	12	10	12	10
25/30	10	8	10	8
35/40	8	6	8	6
45/50	6	4	6	4
60	4	4	4	4
70	4	3	4	3
80	3	2	3	3
90	2	1	3	2
100	1	0	2	1
110	0	001	1	0
125	0	000	1	00
150	00	0000	0	000
175	000		00	0000
200	0000		000	
225			0000	

the source. For this reason, the electrician can run wire and cable under, over, around or through whatever obstacles lay in the path between the circuit source and the outlet. Cable and wire are priced by the foot, however, and any additional footage not planned for can add up fast. The most economical approach to laying out wire runs is to take the most direct route possible and keep the wire (or cable) runs as short as possible.

In new construction, laying out direct, short runs is not too difficult; with open wall, ceiling, and floor cavities, most of the cables proceed directly to their destinations. However, when running cable in existing structures, careful thought must be given to the best and shortest route available from service panel to outlet. Planning the positioning of individual circuits can do much for keeping the total footage of wire required on a job under control. For example, locating the dryer as near to the service panel as is feasible allows the electrician to minimize the amount of heavier, more costly cable needed for the 240-volt circuit required by the dryer. Another benefit of locating the dryer close to the service panel is that it saves the electrician from having to bend and install long runs of conduit to carry the wire.

Use of a junction box allows a considerable savings of money and effort by bringing a single three-conductor cable from the service panel to the box, then splitting off with two-wire cable where the circuits diverge. For indoor jobs, thin-wall electrical metallic tubing (EMT) is generally selected. EMT is normally sold in 10-foot sections of ½-, ¾-, or 1-inch diameter that is shaped with a conduit bender on site.

As with the other construction trades mentioned in this book, the electrical contractor or electrician can benefit greatly from the use of a computer and the software which helps to track material and labor costs.

Using one of the many construction programs to perform the materials take-off list functions, the electrician can have a precise list of what is needed on each job. Combining this list with a program that stores the unit costs for each item and also calculates the total cost per item will bring cost control back in line. Another function of such software is to calculate and track labor costs for multiple jobs simultaneously. Remember, if a two-man crew requires one hour to perform a task, labor is equal to two man hours.

Review questions

1. Every insulated wire (wire size called #14) must have 2.25 cubic inches of space. Any number of bare wires only needs another 2.25 cubic inches of space. Each cable contains two (2) insulated and one (1) bare wire. All boxes are 2 inches wide, 3½ inches high, but some come in different depths. How many cables can be put into a box that is 3 inches deep?
2. Suppose you have a voltage difference of 5 volts. Point A is at 5 volts and Point B is at 0 volts (ground). You want a current to flow between Points A and B and the current needs to be 0.02 amps ($I = 0.02$ amps = 20 mA). Find the value of the resistor.
3. The voltage in a circuit is 115 volts and the resistance is 40 ohms. What is the current?
4. In another circuit, the current is 30 amps and the resistance is 6.5 ohms. A second resistor is added to the circuit giving a total resistance of 15 ohms in the circuit. What is the current after the second resistor has been added?
5. A 15-amp circuit has a resistance of 90 ohms. What will the current be if the resistance is reduced to 60 ohms?

13

Summary of construction math procedures

Many formulas, equations, and mathematical procedures are discussed in this book. Some of these equations require calculating such things as square roots of numbers, which are not the easiest math functions to perform by hand. Fortunately, calculators and computers, which perform these (and many more complicated) math functions, are relatively inexpensive, accurate, reliable, and abundantly available today.

Performing mathematical functions with a calculator

Depending on which of the hundreds of calculators is being used, the keyboard itself will be self-explanatory to an extent. Beyond the obvious addition, multiplication, subtraction and division, keys marked x^2 are used to square numbers (multiply the number by itself). Starting with a blank display, entering a number through the keyboard and then pushing the x^2 button will automatically return the square of the number which was entered. Likewise, a key marked with the symbol $\sqrt{\ }$ represents the mathematical function of determining the square root of a number. A square root of a given number is that number that, when multiplied by itself (squared), equals the given number. For example, 5 is the square root of 25, since $5 \times 5 = 25$. By entering a number on the keyboard of the calculator and then pressing the square root key ($\sqrt{\ }$), one can have the calculator automatically determine and display the square root of the number entered.

Any calculator that performs mathematical functions beyond square roots will, more than likely, come equipped with an owner's manual. Within the manual, exact step-by-step instructions should be listed for each and every function of which the calculator is capable. Most calculators that perform higher-level calculations are equipped with multiple memory storage, allowing the operator to perform calculations and store the results for use in further calculations or for later reference.

Algebraic equations

One of the most useful performance features on calculators today is their capability to compute algebraic equations with the data entered in the same manner as it would be written by hand. For example, an equation that states $A = (B + C)^2$ requires the calculations within parentheses be performed first. On a calculator, the equation would be entered by the following procedure:

1. Press the key for the left parentheses first.
2. Enter the value for *A* followed by the + sign.
3. Enter the value for *B*.
4. Press the key for the right parentheses.

At this point the display on the calculator will be returning the value of (A + B). To complete the equation, two more steps are necessary:

5. Press the x2 key.
6. Press the = key.

The answer presented in the calculator's display will be the value of *A*.

If an equation requires the use of more than one set of parentheses, follow the procedure as described, continuing to enter the data the same way as it would be written. The calculator will keep track of the different values determined by the individual computations and will deliver the final result after the = key has been depressed.

Performing mathematical calculations with a computer

Today's personal computers, whether a desktop model, a lap top model, a PC, or a MacIntosh, is capable of performing hundreds of thousands of mathematical functions per minute; it is what computers do best. The internal "language" all computers use is based on mathematics. Once the appropriate software, such as a spreadsheet program, has been installed on the computer's hard drive (which is the computers major long-term memory center and storage area), the computer will be able to translate the numbers and functions entered through the program into the internal language it uses.

Software

There are virtually thousands of types of software available today. Every arcade-type game designed to be played on a computer for entertainment is actually a computer program. The programs of interest with regard to the type of math discussed in this book are spreadsheet programs, word processing programs, and financial programs. With the use of these three programs, a small contractor can level the playing field with the larger firms, and larger firms can trim some of the fat in their operations.

Each type of program comes complete with instructions on how to be loaded onto the computer and how to be used to perform different functions. As with most situations involving instructions, how easy a program is to use will depend on how complete and how well-written the instructions are. Fortunately, an abundance of books detail how to use most major programs. Some of these books are produced by the program's manufacturer, while others are produced by independent sources in an effort to supplement, compliment, or compete with the manufacturer of the software. In any case, a good bookstore should be able to supply a variety of choices when it comes to instructions for the major categories of business and/or math software programs.

Summary of concepts & formulas

This summary provides a quick reference for concepts and mathematical formulas used elsewhere in this book. For detailed explanations and examples of these concepts and formulas, consult the appropriate chapter.

Rafter size

When determining the length of a roof rafter, view the elements of the roof geometrically. The rafter (or any member forming the longest leg) can also be referred to as the hypotenuse of the right triangle formed by the roofing members.

The algebraic equation used to determine the four elements of roof framing units of measure is:

$$\frac{\text{Total run}}{\text{Total rise}} = \frac{\text{Unit of run}}{\text{Unit of rise}}$$

When dealing with algebraic equations, the goal is to isolate the unknown factor (in this case the X) on one side of the equal to sign (=) with any known, or calculable values on the other side. Once the unknown factor has been isolated, reducing the terms of the equation to their simplest forms will deliver the value of the unknown factor.

Basic formulas and concepts relating to roof framing are given here:

- Any time the total run and total rise have the same value (measurement), the unit of run and unit of rise will also have the same value.
- Multiplying the bridge measure by the total run divided by the unit of run determines the total length of the rafter.
- The most commonly found right triangle with unequal sides has proportions of 18 : 24 : 30 and is most often referred to as a 3 : 4 : 5 right triangle. Any triangle with these proportions is a right triangle.

Estimating

The following principles provide a quick reference to the math involved in estimating for different areas of construction.

Takeoff lists & bills of material

- To calculate materials more precisely (and with less effort), convert the standard lumber sizes and the length in place measurements from a combination of feet and inches into measures of just inches. An 8-foot piece of stock becomes 96 inches, 10 feet becomes 120 inches, and 12 feet converts to 144 inches, while the length in place measurement changes from 1 foot 5 inches to 17 inches. Simple division then shows that dividing a 120-inch length of stock by 17 inches results in seven usable pieces with only 1 inch of waste.
- Material of 1 inch or less (flooring, sheathing, etc.), should have a waste allowance of 20 percent added to their totals, while materials of 2 inches or better should be estimated with a 10-percent waste factor added to their totals.
- To properly estimate labor costs on any construction job, two components must be considered:
  - ~ Manhours required to perform the work.
  - ~ Taxes and insurance costs associated with the total manhours allocated to the job.

labor rate per hour × total number of manhours = total labor cost

Measuring grade

- On level surfaces of equal elevation, grade rod measurements will be the same across all points surveyed from the same instrument location.
- The instrument height will always be greater than the elevation of the existing surface at any point on the site; however, it may be less than or greater than the required plot plan elevation.
- If the instrument height is less than the elevation required by the plot plan for that area, the difference will equal the sum of the grade rod and ground rod, and the area will need to be filled.
- If the instrument height is greater than the required plot plan elevation called for, the difference between the ground rod and the grade rod will equal the amount of cut or fill required.

- Determine the depth of the excavation by adding the vertical distance between the finished basement floor surface and the first floor finished planes, plus the 9 inches below the finished basement floor surface.

Interest rates

- *Periodic rate.* The rate of interest applied to the loan one period at a time. Periods can be monthly, quarterly, semi-annually, or annually.
- *Annual nominal rate.* The periodic rate multiplied by the number of periods in a year. If the monthly periodic rate is ½ percent, the annual nominal rate would be 6 percent (½ × 12 = 6).
- *Annual effective rate.* The annual rate that takes compounding into consideration and is today usually referred to as the Annual Percentage Rate (APR).

Compounding refers to the practice of periodically assessing interest on the unpaid balance of a loan. To determine different aspects of compound interest bearing loans, the following formula is used:

$$A = P \times (1 + i)^n$$

where

A = the total repayment amount of the loan
P = the principal (the amount borrowed)
i = the periodic rate of interest
n = the number of periods in the term of the loan (expressed as an exponent).

Interest equals principal times rate times time.

Proportional estimating of concrete

- The $3/2_s$ rule refers to an assumption stating that the combined components of a concrete mix, for any given volume of concrete, will be $3/2_s$ (1½ times) the volume of the concrete pour.

- The ratio of the dry components is usually taken as 1 : 2 : 3 (read: 1 to 2 to 3) referring to 1 part cement, 2 parts fine aggregate, and 3 parts coarse aggregate.
- Adding the elements of the ratio (1 + 2 + 3) gives six parts to the mix:
  - ~ 1 part cement = ⅙ of the total volume
  - ~ 2 parts fine aggregate = 2/6 of the total volume
  - ~ 3 parts coarse aggregate = 3/6 of the total volume

Brick masonry

- For a planned wall with standard brick and ¼-inch mortar joints, multiplying the number of square feet to be bricked by 7 will result in the number of bricks needed for the construction of a 4-inch thick wall (which is the thickness of one brick, plus its joints).
- If the area to be bricked requires an 8-inch thick wall, multiply by 14 instead of 7.
- A 12-inch thick wall needs to be multiplied by 21 to arrive at the necessary number of bricks.

Determining beam size

1. From one side of the beam, measure the joist span (which runs from the middle of the beam to the nearest joist support).
2. Determine the measurement of the corresponding span on the other side of the beam and add the two measurements together.
3. If the joists are butted or lapped over the beam, multiply the total measurement from #2 by ½; if the joists are continuous, multiply by ⅝ instead of ½.
4. Add the dead and live load figures for each floor; the dead load figures for each floor partition, and the dead and live load figures for the roof (taken from the plans). The total of these figures equal the combined square foot floor load that the beam is to carry.
5. Multiply the combined square foot floor load figure by the half width to obtain the amount of load on the beam per linear foot.

6. Multiply the load on the beam per linear foot by the distance between beam supports to obtain the total load on the beam.

Determining the amount of studs needed

1. Sum the total footage from the lengths of all walls and partitions and multiply by ¾ (or 75 percent). For example, 200 feet of walls and partitions × ¾ = 150 feet.
2. Count the number of wall and partition segments and multiply by the length of one stud (8 feet). Assuming there are a total of 10 walls and partitions, 10 × 8 feet = 80 feet.
3. Count the number of times an exterior wall is intersected by a partition, and multiply this number by the length of one stud (8 feet). If this situation occurs five times, 5 × 8 feet = 40 feet.
4. Count the number of times one partition intersect another partition and multiply this number by the length of one stud. If there are four instances where one partition intersects another, then 4 × 8 feet = 32 feet.
5. Count the number of openings in all walls and partitions. Double this figure and multiply by the length of one stud.
6. To allow for plates and caps, multiply the answer from the first step by 3, then divide that answer by the length of one stud (8 feet).
7. Total the answers from each step.
8. Divide the total by the length of one stud.
9. Add 5 percent to the total for waste.

Determining roof rise & span

The span of a roof is the distance, measured horizontally, from one outside plate (rafter seat) to the outside plate directly opposite. The total rise of a roof is the distance, measured vertically from one outside plate to the center of the ridge line.

The Pythagorean Theorem

The Pythagorean Theorem states: In a right triangle, if A, B and C are the lengths of the sides of a triangle, where C is the length of the hypotenuse, then $A^2 + B^2 = C^2$ (A squared plus B squared equals C squared).

Determining valley rafter length

Once the bridge measure has been derived, multiplying it by ½ the span of the roof for which the valley is intended will provide the length of the valley rafter in inches; divide by 12 to arrive at the length expressed in feet.

Formulas for calculating roof area

- Figuring the area of a *shed roof* requires multiplying line length from the eave edge to the ridge, by the span.
- A *gable roof* requires the multiplying of the distance from the eave edge to the ridge by twice the length of the ridge line to determine the total area of the roof.
- If sheathing boards are to be used, the quantity of material (in board feet) is the same as the total area.
- If a *hip roof* is on a square building, the geometric formula for determining the area of a triangle becomes very useful. Separating the roof into 4 equal triangles, the area of one of the four triangles is calculated with the formula

  $$A = \frac{1}{2} bh$$

  which reads Area = half the base of the triangle times its height. The area of one triangle is 150 square feet. Since the roof comprises four triangles of equal dimensions, multiplying the area of one triangle by 4 will equal the total area of the roof.
- To calculate the area of a hip roof on a rectangular structure:
  1. Multiply the length of the eave edge along the short side of the rectangle (A) by the height of the triangle (B) formed by the roof on that side (measuring from the eave edge to the ridge).

2. Determine the length of the eave edge along the long side of the rectangle (C). Add this figure to the length of the ridge line (E) and multiply the sum by the height of the line measure from eave edge to ridge line (D).
3. Adding the results of step No. 1 and step No. 2 will equal the total area of the roof.

Area equals A times B plus the product of D times the sum of C plus E: Area $= (A \times B) + D(C + E)$.

Fahrenheit & Centigrade conversions

In degrees Fahrenheit, water freezes at 32 degrees and boils at 212 degrees. According to the Centigrade scale, water freezes at 0 degrees and boils at 100 degrees.

To convert from Fahrenheit to Centigrade, subtract 32 from the Fahrenheit temperature and divide the result by 1.8. To convert from Centigrade to Fahrenheit, multiply the Centigrade temperature by 1.8 and add 32.

Calculating the required ventilation

The area to be vented (the attic) should be measured to determine its square footage. This holds true whether passive or power ventilation is to be used. With passive systems, the square foot figure for the attic is then divided by 150, if there is no vapor barrier facing the living space or 300 if there is a vapor barrier in place. This calculation will yield the required number of square feet of ventilation. Half the number of square feet dedicated to ventilation must be designed for intake with the other half designed for exhaust.

Water requirements

The total gallons per hour of demand is equal to the number of fixtures multiplied by 60. Expressed mathematically as:

$$\text{gph} = \text{fixtures} \times 60$$

Pressure loss

For every 2.3 feet of vertical run, there is a drop of 1 pound per square inch of pressure.

2.3 feet = 1 psi

Measuring pipe

The equation for the center-to-center method, as illustrated in Fig. 11-2 is as follows:

$$D = A - 2B + 2C$$

where *D* stands for distance (the value being calculated). The *A* represents the measured distance from the exact center of one fitting to the exact center of the fitting on the opposing end of the pipe. The *B* represents the distance from the face of one fitting to the center of that same fitting. Because the pipe will not extend into the fitting as far as the center of that fitting, the length of *B* must be subtracted from both ends of the pipe, which is why the formula calls for the subtraction of 2*B*. The *C* in the formula represents the actual length of pipe that will extend into the fittings. Here again, since this occurs on both ends of the pipe, the formula stipulates the addition of 2*C*. *AC* equals the square root of $(AB^2 + BC^2)$, where *AC* is the length of the pipe (the value being calculated), and *AB* and *BC* equal the offsets.

Applying math to electricity

amps = watts divided by volts

Ohm's Law

$I = E/R$, where *I* represents the flow of current expressed in amperes, *E* stands for the amount of voltage (in volts), and *R* represents the amount of resistance present in the circuit.

Transposition

Rearranging an algebraic equation is accomplished through a method called *transposition*. In common-sense terms, when an equation is set up to calculate the value of one of the elements in the equation (such as I in the example of Ohm's Law above), the unknown value is isolated on one side of the equal sign in the equation. If one of the other elements is the value to be solved for, then that element would need to be isolated. To accomplish this, start with the original equation:

$$I = \frac{E}{R}$$

To isolate E, something must be done to remove R from the side E is on. Because the original equation calls for E to be divided by R, multiplying E by R removes R from that side of the equation, leaving E all by itself. As with all algebraic equations, whatever is done to one side of the equation must be done equally to the other side of the equation, so I must also be multiplied by R. This results in the equation becoming IR (or $I \times R$) instead of just I, and it will read as follows:

$$IR = E$$

Circular mils

Converting the area of a circle into circular mils (CM) is done by taking the diameter of the circle (in mils) and multiplying that figure by itself (squaring the figure). The mathematical equation for this would appear as

$$A = D^2$$

where A is the area to be determined in circular mils and D is equal to the diameter of the circle measured in mils.

Determining resistivity

The formula for determining the resistivity of a conductor is

$$R = \frac{KL}{A}$$

where R equals the material's resistivity expressed in ohms, K represents the specific resistance of the material, and L is the length of the conductor while A equals the cross-sectional area of the conductor. By substituting the values of R (resistance), E (voltage), and I (current) from Ohm's law, the formula for determining resistance will provide the means to determine the size of wire required for a given set of circumstances.

$$d^2 = \frac{L \times K}{\left(\frac{E}{I}\right)} \text{ or}$$

$$d^2 = \frac{L \times K \times I}{E}$$

where d^2 is the cross-sectional area of the wire (the size or gauge wire to be used), L is the length of the wire, K is the specific resistance of the material being used as a conductor, I represents the amount of current flowing through the circuit, and E is the amount of voltage being fed into the circuit.

Appendix A

Metric conversions for the building trades

Property	**To convert from**	**Symbol**
Application rate	U.S. gallon per square	gal (U.S.)/100 ft^2
	U.K. gallon per square	gal (U.K.)/100 ft^2
Area	square inch	$in.^2$
	square foot	ft^2
	square	100 ft^2
Breaking strength	pound force per inch width	lbf/in.
Coverage	square foot per U.S. gallon	ft^2/gal
	square foot per U.K. gallon	ft^2/gal
Density, or mass per unit volume	pound per cubic foot	lb/ft^3
Energy or work	kilowatt-hour	k Wh
	British thermal unit	
Flow, or volume per unit time	U.S. gallon per minute	gpm
	U.K. gallon per minute	gpm
Force	pound force	lbf
	kilogram force	kgf
Heat flow	thermal conductance, C	Btu/h • ft^2 • °F
	thermal conductivity, k	Btu • in./h • ft^2 • °F
Incline	inch per foot	in./ft
Length, width thickness	mil	0.001 in.
	inch (up to ~48 in.)	in.
	foot (~4 ft and above)	ft
Mass (weight)	ounce	oz
	pound	lb
	short ton	2000 lb
Mass per unit area	pound per square foot	lb/ft^2
	pound per square foot	lb/ft^2
	pound per square foot	lb/1000 ft^2
	ounce per square yard	oz/yd^2

to	Symbol	Multiply by	Remarks
litre per square metre	litre/m^2	0.4075	= 0.4075 mm thick
litre per square metre	litre/m^2	0.4893	=0.4893 mm thick
square millimeter	mm^2	645.2	1,000,000 mm^2 =1 m^2
square metre	m^2	0.092 90	
square metre	m2	9.290	
kilonewton per metre width	kN/m	0.175	
square metre per litre	m^2/litre	0.024 54	
square metre per litre	m^2/litre	0.020 44	
kilogram per cubic metre	kg/m^3	16.02	water = 1000 kg/m^3
megajoule	MJ	3.600*	J = W • s = N • m
joule	J	1055	
cubic centimetre per second	cm^3/s	63.09	or 0.0631 litre/s
cubic centimetre per second	cm^3/s	75.77	or 0.0758 litre/s
newton	N	4.448	N = kg • m/s^2
newton	N	9.807	
watt per square metre kelvin	W/m^2 • K	5.678	
watt per metre kelvin	W/m • K	0.1442	
percent	%	8.333	3 in./ft = 25%
micrometre	μm	25.40*	1000 μm = 1 mm
millimetre	mm	25.40*	1000 mm = 1 m
metre	m	0.3048*	
gram	g	28.35	1000 g = 1 kg
kilogram	kg	0.4536	1000 kg = 1Mg
megagram	Mg	0.9072	
kilogram per square metre	kg/m^2	4.882	
gram per square metre	g/m^2	4882	
gram per square metre	g/m^2	48.82	
gram per square metre	g/m^2	33.91	

Property	To convert from	Symbol
Permeability at 23°C	perm inch	grain • in./ft^2 • h • in. Hg
Permeance at 23°C	perm	grain/ft^2 • h • in. Hg
Power	horsepower	hp
Pressure or stress	pound force per square inch	lbf/in.^{2} or psi
	pound force per square foot	lbf/ft^2 or psf
Temperature	degree Fahrenheit	°F
	degree Celsius	°C
Thread count (fabric)	threads per inch width	threads/in.
Velocity (speed)	foot per minute	ft/min or fpm
	mile per hour	mile/h or mph
Volume	U.S. gallon	gal (U.S.)
	U.K. gallon	gal (U.K.)
	cubic foot	ft^3
	cubic yard	yd^3

to	Symbol	Multiply by	Remarks
nanogram/pascal second metre	ng/Pa • s • m^2	1.459	ng = 10^{12} kg
nanogram/pascal second square metre	ng/Pa • s • m	57.45 64 mg	lgrain =
watt	W	746	W = N • m/s J/s
kilopascal	kPa	6.895	Pa = N/m^2
pascal	Pa	47.88	
degree Celsius	°C	$(t_F - 32)/1.8$*	32°F ≈ 0°C
kelvin	K	$t_c + 273.15$*	273.15K ≈ 0°
threads per centimetre width	threads/cm	0.394	
metre per second	m/s	0.005 080*	
kilometre per hour	km/h	1.609	
cubic metre or 3.785 litres	m3	0.003 785	
cubic metre or 4.546 litres	m3	0.004 546	
cubic metre	m3	0.028 32	
cubic metre	m3	0.764 6	

*Exact conversion factor

Appendix B

Decimal equivalents of common fractions

Common Fraction	Decimal Equivalent	Common Fraction	Decimal Equivalent
1/2	0.50	11/12	.9166
1/3	0.3333	1/16	.0625
2/3	0.6666	3/16	.1875
1/4	0.25	5/16	.3125
3/4	0.75	7/16	.4375
1/5	0.20	9/16	.5625
2/5	0.40	11/16	.6875
3/5	0.60	15/16	.9375
4/5	0.80	1/24	.0417
1/6	0.1666	5/24	.2083
5/6	0.8333	7/24	.2917
1/7	0.1428	9/24	.3750
2/7	0.2857	11/24	.4583
3/7	0.4286	13/24	.5417
4/7	0.5714	17/24	.7083
5/7	0.7143	19/24	.7917
6/7	0.8571	23/24	.9583
1/8	0.1250	1/32	.0313
3/8	0.3750	3/32	.0938
5/8	0.6250	5/32	.1563
7/8	0.8750	7/32	.2188
1/12	0.0834	9/32	.2813
5/12	0.4166	11/32	.3437
7/12	0.5833	13/32	.4063
15/32	0.4688	17/32	.5313
19/32	0.5938	21/32	.6563
23/32	0.7188	25/32	.7813
27/32	0.8438	29/32	.9063
31/32	0.9688		

Note: The decimal equivalents in this table are rounded to 4 decimal places. By dividing the numerator by the denominator of any of the common fractions the decimal equivalent can be extended to as many decimal places as needed.

Appendix C

Answers to review questions

Chapter 2

1. $\frac{1}{2} + \frac{1}{3}$ cannot be combined because they do not have the same denominator. Fractions $\frac{1}{2}$ and $\frac{1}{3}$ have the least common denominator (LCD) of 6. Therefore, fraction $\frac{1}{2}$ has to be multiplied by $\frac{3}{3}$ to make $\frac{3}{6}$ and fraction $\frac{1}{3}$ has to be multiplied by $\frac{2}{2}$ to make $\frac{2}{6}$.

$$\frac{1}{2} \times \frac{3}{3} = \frac{3}{6}$$

$$\frac{1}{3} \times \frac{2}{2} = \frac{2}{6}$$

Now add the fractions $\frac{3}{6}$ and $\frac{2}{6}$ because they have the same common denominator.

$$\frac{3}{6} + \frac{2}{6} = \frac{5}{6}$$

2. As in question 1 the fractions $\frac{1}{2} - \frac{1}{3}$ cannot be subtracted because they do not have the same common denominator. Fractions $\frac{1}{2}$ and $\frac{1}{3}$ have the least common denominator (LCD) of 6. Therefore, just as in question 1 the fraction $\frac{1}{2}$ has to be multiplied by $\frac{3}{3}$ to make it $\frac{3}{6}$ and the fraction $\frac{1}{3}$ has to be multiplied by $\frac{2}{2}$ to make it $\frac{2}{6}$. Now subtract $\frac{2}{6}$ from $\frac{3}{6}$ as they now have the same common denominator.

$$\frac{3}{6} - \frac{2}{6} = \frac{1}{6}$$

3. When we divide a whole number by a fraction, we multiply the whole number with the ***inverse*** of the fraction. Substituting the numerator for the denominator and the denominator for the numerator creates the inverse of a fraction. In this question, the ***inverse*** of $\frac{1}{3}$ is $\frac{3}{1}$. And the product of 2 and 3 is 6. There are 6 thirds ($\frac{1}{3}$s) in the whole number 2.
4. When dividing fractions, multiply the first fraction with the ***inverse*** of the second fraction. In this question, the inverse of $\frac{1}{3}$ is $\frac{3}{1}$.

$$\frac{1}{2} \times \frac{3}{1} = \frac{3}{2} \text{ or } 1\frac{1}{2}$$

5.

$$16 : 8 = x : 5$$

$$\_\_ = \_\_$$

$$8 \quad 5$$

$$x \times 8 = 16 \times 5$$

$$x = 10$$

Chapter 3

1. This answer is obtained by knowing about proportions and how they are used. You can set up proportions by using ratios. Remember, ratios are comparing ***similar*** things. The first ratio is comparing employees and the second is comparing tasks.

$$2 : 5 = 20 : x$$
$$2 \times x = 5 \times 20$$
$$x = 50 \text{ tasks}$$

2. Estimated labor per house would be $12,820 (2000 SF × $6.41 labor per SF). For both houses multiply by 2 = $25,640.
3. $5.98 per square foot ($20,930 ÷ 3500 SF = $5.98).
4.
   - Track operating and maintenance costs,
   - allocate their cost to jobs, and
   - schedule maintenance.
5. 2000 SF ÷ 15SF/hour =133.333 hours × 2 men = 266.666 man hours.

Chapter 4

1a. What is the minimum price the land company must receive, on a square footage basis, for the land in this subdivision?
Cost of land:

76.80 acres × $20,000/acre = $1,536,000

Total cost to develop the subdivision equals cost of land plus cost of engineering and improvements:

$1,536,000 + $1,250,000 = $2,786,000

The total minimum price equals total cost to develop the subdivision times 250%:

$2,786,000 × 2.50 = $6,965,000

Total square footage in the development equals total acres times the square footage per acre:

76.80 acres × 43,560 sq ft/acre = 3,345,408 square feet

Total sellable square footage equals total square footage in the development less square footage to be dedicated as roadways, sidewalks, and median strips:

3,345,408 sq ft − 415,000 sq ft = 2,930,408 sq ft

The minimum sales price per square foot equals the total minimum sales price for the subdivision divided by the sellable square footage in the development:

$6,965,000.00 2,930,408 sq ft = $2.3768021 or $2.38

b. Does it make economic sense for the land company to purchase this property?

- Yes, this land can be sold for $3.50 per square foot and $2.38 per square foot will ensure an adequate profit. If the land is sold for $3.50 per square foot, Eagle Land Company would receive an additional profit of $1.12 per square foot above the amount it deems to be an adequate profit.
- At 2,930,408 square feet of sellable land in the entire development, the additional $1.12 per square foot would result in an extra $3,282,056.90 in profit on property that cost $2,786,000 to develop.

2a. Your total estimated project costs are $210,000. See Table A4-1.

Table A4-1

Personnel	$100,000
Materials and supplies	$50,000
Permitting & related costs	$10,000
Overhead	$20,000
Taxes	$30,000
TOTAL	**$210,000**

Your total estimated project costs with contingency for cost overruns are $241,500. See Table A4-2.

Table A4-2

Total estimated project costs	$210,000
Contingency	× 15 %
Total additional contingency amount	$31,500
TOTAL w/contingency (cost + contingency)	**$241,500**

Your total bid to the customer should be $289,800. See Table A4-3.

Table A4-3

Total estimated project costs with contingency	$241,500
Profit	× 20%
Total additional profit	$48,300
Total bid (costs + contingency + profit)	**$289,800**

Table A4-4

Total estimated project costs	$210,000
Total bid with profit	$289,800
Total cost with no overruns	−$210,000
Total profit with no overruns	$79,800

b. Your percent profit if no cost overruns are experienced would be 27%.

Profit total bid = % profit

3. Find the starting point marked (D) (See Fig. A4-1). Measure the width of the house to find the distance (DE).
Measure the length to find distance (EF).
Using the Pythagorean theorem:

$$\mathbf{DF = (FE)^2 + (DE)^2}$$

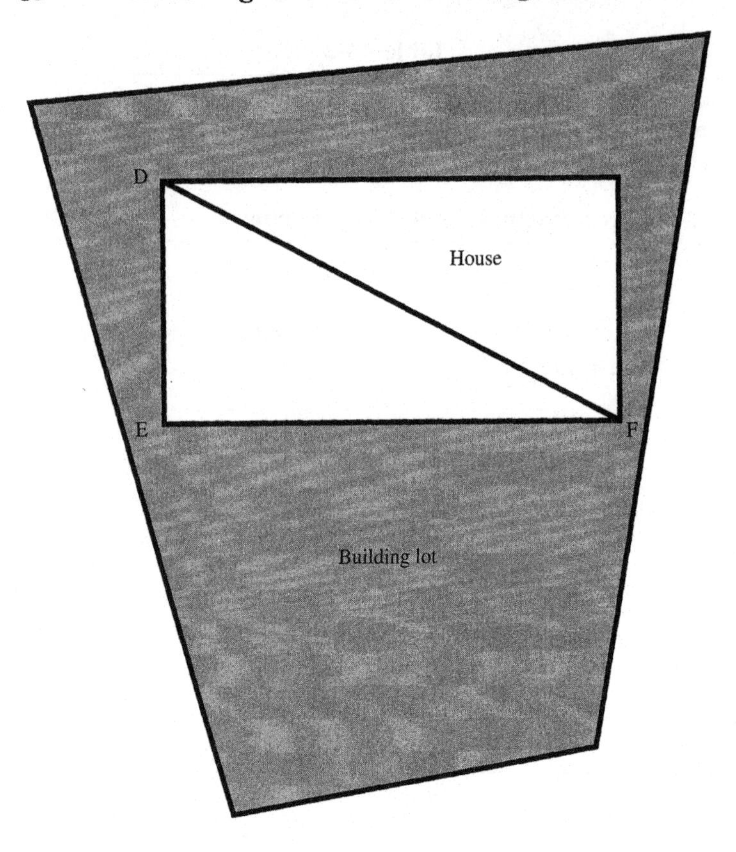

A4-1

Chapter 5

1a. 1 cubic yard = 27 cubic feet. 4 inches = $\frac{4}{12}$ of a foot. Volume of driveway: 135 ft × 15 ft × $\frac{4}{12}$ ft = 675 cu ft. 675 ft^3 ÷ 27 ft^3 = 25 yd^3.

$$25\ yd^3 \times \$55 \text{ per } yd^3 = \$1{,}375$$

b. If the bill is paid in cash, take 15% of the bill or $206.25 as a deduction. Final bill = $1,168.75

2. 3 inches ÷ 1 foot (12 inches) = ¼ foot or .25 ft. 144 tons ÷ 2000 ft^3 = 0.72 tons of asphalt per cubic foot.

length • width • depth • weight per ft^3 = tons of asphalt
5280 ft × 26 ft × .25 ft × .072 tons = 2471 tons of asphalt needed.

3.
- 1 cubic yard = 27 cubic feet (ft^3)
- 1 cubic foot = 3 ft^2
- 1 cubic yard = 81 ft^2
- 4 inch = ⅓ foot
- 4 ft × 16 ft × ⅓ ft = 21.3 ft^3

a. 21.3 ft^3 × 1 yd^3
27 ft^3 = 0.79 cubic yards needed
b. .79 cubic yards × $70.00 per cubic yard = $55.30

4. Use the Pythagorean Theorem.

$$44^2 + 68^2 = X^2$$
$$\sqrt{6560} = X$$
$$80.99 = X$$
$$81 \text{ ft} = X$$

Should this number be equal to the length when measured, you will have a 90-degree angle.

5. Number of square feet = 68 (length) × 6 (height) = 408 ft^2. One square foot of 4 inch thick wall requires 7 standard brick with ¼ inch mortar; 408 × 7 = 2856 bricks.

Chapter 6

1. The window header will carry ½ of the roof load or a "tributary" load based on 20′0″.

Roof live load = 30#/ft^2 × 20′ = 600#/ft
Roof dead load = 15#/ft^2 × 20′ = 300#/ft
Total uniform load = 900#/ft.

2. From Table 6-8, choose the span equal to or longer than the required span (10′0″ span is chosen).
Read down the column to a load bearing capacity equal to or larger than the required capacity. Keep in

mind the beam must be able to carry its own load as well and that you will need to add on the weight per lineal foot of the determined beam.

The 4½″ × 11½″ Glu/Lam is chosen and it is noted that the weight of the beam is 12#/ft. The total load required is 912#/ft, and the chosen member over a 10′0″ span will carry 1036#/ft.

3. 17% of 1720 ft^2 = 292.4 ft^2; Each window is 3′ × 5′ or 15 ft^2; 292.4 divided by 15 = 19.49. Therefore, Mr. and Mrs. Johns can have a maximum of 19 windows.

4a. $$24 \text{ ft} \times 18 \text{ ft} = 432 \text{ ft}^2$$

432 ft^2 × \$9 per ft^2 (\$7 per ft^2 for materials plus \$2 per ft^2 for installation) = \$3,888 for hardwood flooring.

b. A standard sheet of plywood measures 4 ft × 8 ft = 32 ft^2
Divide a 432 ft^2 room by 32 ft^2 = 13.5 sheets of plywood.

5. $12 : 52 = 18 : x$
$x \times 12 = 52 \times 18$
$X = 78$ pounds

Chapter 7

1. Figuring the freight:
The truck needs \$1.25 per loaded mile for a total of 650 miles. \$812.50 ÷ 24 (thousands of board feet) = \$33.854 or approximately \$34.00 per thousand board feet.
   a. Shipment composition:
   Knowing how many lengths the customer wants it is necessary to keep a running total of the lineal footage, then multiply the total by the factor for the board size to get the number of board feet.
   2 × 4s:

$$(200 \times 8) + (200 \times 10) + (200 \times 12) + (400 \times 14) + (400 \times 16) = 18{,}000$$

18,000 ft × .6667 (factor for 2 × 4S) = 12,000 board feet.

2 × 10s:

$$(80 \times 8) + (80 \times 10) + (80 \times 12) + (160 \times 14) + (160 \times 16) = 7200 \text{ feet.}$$

7200 feet × 1.6667 (factor for 2 × 10s) = 12,000 board feet.

From this, it is determined that what the customer wants will fit on one truck.

b. Quoting a price:
The price is determined by the cost of the material at the source, plus the freight, plus the broker's profit.
2 × 4s:

$380.00 ÷ $34.00 = $414.00 + 4% =
$431.00 (430.56) per thousand board feet.

2 × 10s:

$440.00 + $34.00 = $474.00 +4% =
$493.00 (492.96) per thousand board feet.

$431.00 + $493.00 = $924.00 total price.

2. To find how many sheets will be necessary for the 4 walls, divide the area of the walls by the area of a plywood sheet (not allowing for doors or windows).

$$\frac{[2 \text{ walls } (26' \times 8')] + [2 \text{ walls } (22' \times 8')]}{4' \times 8' \text{ plywood sheets}}$$

$$(2 \times 208 \text{ ft}^2) + (2 \times 176 \text{ ft}^2)$$

divided by 32 ft^2 = (416 ft^2 + 352^2)/32 ft^2

or

768 ft^2/32 ft^2 = 24 sheets of plywood

3. Height of pt M = 17′ 2-⅞″ + 14″ − 13″ = 16′ 5′⅞″;
Pitch of line LM = 5″ in 12″ (rise/run);
Total run = 20′ − 3½″ = 19′ 8½″ or 19.708 ft (run);
19.708 ft (run) × 5″ (rise/feet) = 98.54 inches (rise nm);
Height of pt L = M − rise nm = (16′ 5 ⅞″) −
(8′ 3 ⅞″) = 8′ 3 5⁄16 = height of wall A.

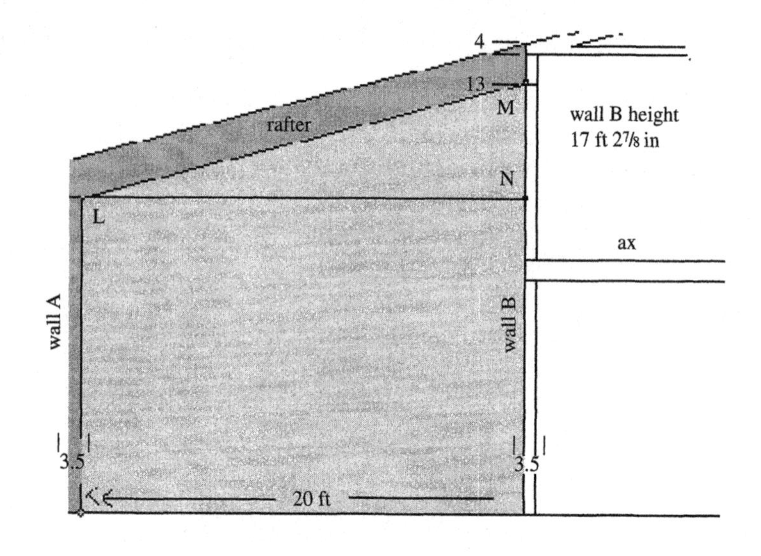

Chapter 8

1. Length of LM $= \sqrt{(8.21\text{ ft})^2 + (19.708\text{ ft})^2} = 21'\ 4\ 3/16''$
2. Consider the end view of the roof (Figure A8-2) as two right triangles, each with a base of $12' + 2' = 14'$

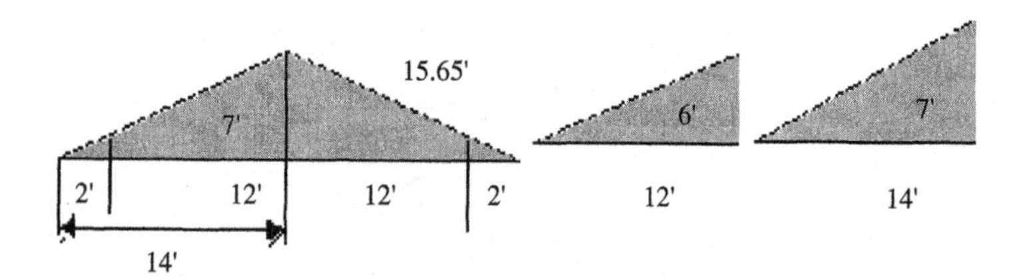

A8-2

With a slope of 6″ in 12,″ the height of the triangle will be 7.′

$$6''/12'' = x''/168'' = 7'/14'$$

$$7^2 + 14^2 = 245\text{ ft}^2 = 15.65\text{ ft}^2$$

Area of half the roof $= 15.65' \times (24' + 4') = 438.2\text{ ft}^2$
Area of the roof $= 438.2 \times 2 = 876.4\text{ ft}^2$

The number of shingle bundles required = the area of the roof divided by 33.33 (⅓).

$$876.4 \div 33.33 = 26.29$$

or 27 bundles.

3. To find how many sheets will be necessary for the roof, divide the total area of the roof as seen in Figure A8-3 (2 equal sides) by the area of a plywood sheet (plywood sheet = 4′ × 8′).

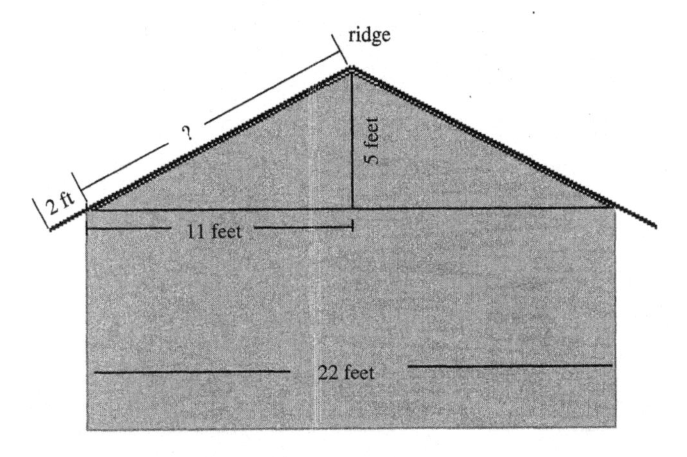

A8-3

The ridge of the roof is 26 feet and the overhand is 2 feet. The height of the roof is 5 feet. Find the area of each side of the roof by calculating the length from the ridge to the edge ($a^2 + b^2 = c^2$), adding the overhang, and multiplying the total length by the width (ridge).

$$5^2 + 11^2 = c^2$$

$$25 + 121 = 146$$

or approximately 12 feet.
Roof area: (2′ + 12′) × 26′ = 364 ft^2 × 2 (both sides) = 728 ft^2
720 ft^2/32 ft ft^2 = 22.75 or 23 sheets of plywood for the roof.

4. 20 ft × 40 ft = 800 ft² × 2 = 1600 ft²
   12 ft² coverage/pkg = 133.3 or 134 packages of shingles.
   a. 134 packages × $7 per package = $938 for shingles.
      134 packages × $12 = $1608 for labor & materials
      $938 + $1608 = $2546 for the installed roof.

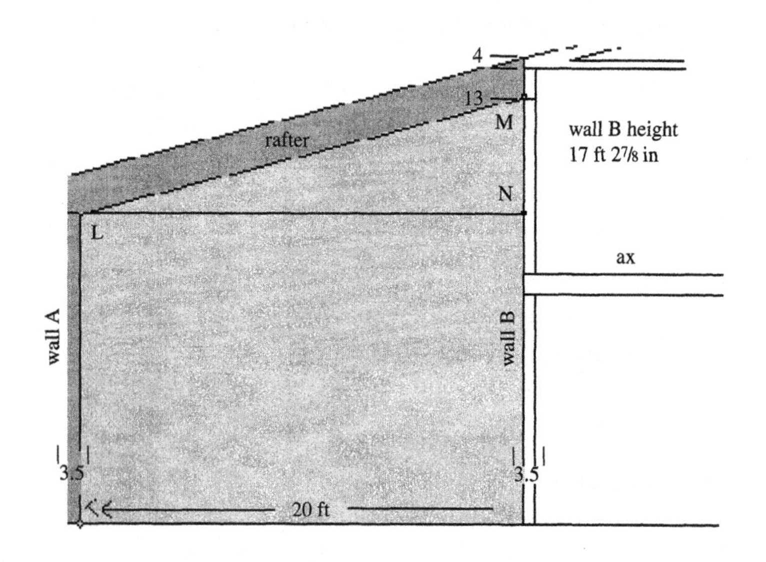

Chapter 9

1. 20 feet × 12 inches/foot = 240 inches
   240 inches divided by 16 inch centers = 15 spaces + 1 = 16 (8-foot) 2 × 4s.
2. 20 feet divided by 4-foot sheets = 5 sheets of sheetrock.
3. Figure the cost for the material in board footage delivered to the site in Chicago. (Cost at the mill per thousand board feet) × (total amount of material in thousands).

   $530 × 98 = $51,940 (cost for material)

   Add the freight total to the cost of material to get the total cost.

$51,940 + $4,420 = $56,360 (total cost with freight)

Break down the total cost into the cost-per-piece for the 5 lengths of 1″ × 6″ boards. Total delivered cost/total board footage = cost for 1 board foot, the equivalent of one 1″ × 12″ linear foot

$56,360/98,000 = $.575 (per 1 board foot, 12″ board)

Next, figure the cost per linear foot for a 1″ × 6″ board $.575 × [(1″ × 6″) ÷ (1″ × 12″)] = $0.2875 (price per linear foot times a 6″ board).
This cost is then multiplied by the length of each board in order to figure the costs per piece of lumber:

$.2875 × 8′ = $2.30 per piece of 1″ × 6″ 8-foot-long board

$.2875 × 10′ = $2.88 per piece of 1″ × 6″ 10-foot-long board

$.2875 × 12′ = $3.45 per piece of 1″ × 6″ 12-foot-long board

$.2875 × 14′ = $4.03 per piece of 1″ × 6″ 14-foot-long board

$.2875 × 16′ = $4.60 per piece of 1″ × 6″ 16-foot-long board

4. 24 ft × 18 ft = 432 sq ft of carpet needed; $13 (carpet) + $5 (installation) = $18/sq yd installed carpet; 432 sq ft/9 sq ft = 48 sq yd × $18/sq yd = $864 for carpet.
5. Area of walls ft² covered by one gallon = 2(L × H) + 2(W × H)

   300 = gallons needed = 2(63 × 10) + 2(47 × 10)

   300 = 7.3 gallons of paint

Chapter 10

1. Yes. The 60 hp fan must meet the part-load efficiency requirement. However, the return fan will normally be exempt because it is smaller than 25 hp.
2. No. IPLVs are determined by measuring performance at steady-state part-load conditions. If the equipment cannot operate at that condition without cycling, its steady-state performance cannot be measured. Thus for a single speed compressor with no cylinder unloading, IPLV requirements do not apply.
3. No. The VAV duct mains supply cooled air and are located over interior spaces that also require cooling.

Heat gains or losses from the ductwork will therefore not affect energy usage. Zone ducts also need not be insulated since they are located in the spaces served.

4. The unit capacities should match the design loads, but can include pick-up, pull-down, and safety factors per 9.3(a)(8) and 9.3(a)(9). These contingency loads are 30% (heating pick-up), 10% (cooling pull-down), and 10% (safety), respectively. The resulting coil capacities are given as follows:

$$\text{Heating capacity} = 360{,}000 \times 1.1 \times 1.3 = 514{,}800 \text{ Btu/h}$$
$$\text{Cooling capacity} = 36 \times 1.1 \times 1.1 = 43.56 \text{ tons}$$

Note that pick-up and pull-down loads are only allowed on systems that are setback, setup, or shutoff.

5. No. If the installed airflow is to be 1.1 cfm/ft^2 $\times$ 1.2 or 1.32 cfm/ft^2, the model must be re-run with this installed value.

Chapter 11

1. $L = R \times D \times 0.01745$

Length of the bend:

$$\text{Length A} = 24 \times 135 \times 0.01745 = 56.538 \text{ inches}$$
$$\text{Length B} = 24 \times 270 \times 0.01745 = 113.076 \text{ inches}$$
$$\text{Length C} = 24 \times 135 \times 0.01745 = 56.538 \text{ inches}$$
$$56.538\ (A) + 113.076\ (B) + 56.538\ (C) = 226.152 \text{ inches}$$

Length of the pipe:

$$L + 2T = 226.152 + 20 = 246.152 \text{ inches}$$

2. $G = P \times D \times D \times L \times 5.875$ (formula when measurements are in feet);
$G = P \times D \times D \times L \times .0034$ (formula when measurements are in inches);
$.7360 \times 4 \times 4 \times 10 \times 5.875 = 691.840$ gallons of water.
3. $236 : 2 = 4543 : X$
4. A difference in height of 30 feet reduces the available pressure by 13 psi. Multiply the elevation of the highest

fixture in feet, by the factor 0.434, the pressure exerted by a 1-foot column of water ($30 \times 0.434 = 13.02$).

5. GPM = 10
   Internal area = .304 (see note below)
   $0.3208 \times \text{GPM/Internal area} = .3208 \times 10 \times .304 =$ 10.55 feet per second.

   Note: The outside diameter of pipe remains the same regardless of the thickness of the pipe. A heavy-duty pipe has a thicker wall than a standard-duty pipe, so the internal diameter of the heavy-duty pipe is smaller than the internal diameter of a standard-duty pipe. The wall thickness and internal diameter of pipes can be found on readily available charts.

Chapter 12

1. Box volume is $3'' \times 3\frac{1}{2}'' \times 2''$ or 21 in^3
   Space needed for insulated wires:

   $$21\ in^3 - 2.25\ in^3 = 18.75\ in^3$$

   (all bare wires need only 2.25 in^3).
   Space needed for each cable that contains two insulated wires:

   $$2.25\ in^3 \times 2 = 4.50\ in^3$$

   Maximum number of cables allowed:

   $$18.75\ in^3 \div 4.50\ in^3 = 4.166\ .\ .\ .$$

   Therefore, 4 cables can be put into the box.
2. Resistance = Difference in Voltage ÷ Current or
   $R = DV \div I$
   $DV \div I = 5 \div 0.02 = 250$ ohms.
   This means that putting a resistor with a value of 250 ohms between Points A and B will make a current flow from Point A to Point B and the current will be 0.02 amps (20 mA).
3. 2.875 amps
4. 13 amps: 30 amps × 6.5 ohms = 195 volts. Then, 195 volts ÷ 15 ohms is the current desired (13 amps).

Or, 30 amps × 6.5 ohms = I amps × 15 ohms. Then, cross multiply to get 30 × 6.5 ÷ 15 = 13 amps.

5. 22.5 amps: 15 amps × 90 ohms = 1350 volts. Then, 1350 volts ÷ 60 ohms is the desired current (22.5 amps). Or, 15 amps × 90 ohms = I amps × 60 ohms. Then, cross multiply to get 15 × 90 ÷ 60 = 22.5 amps.

Glossary

abrasion resistance The ability of a material to resist mechanical action such as foot traffic and wind-blown particles, which tend progressively to remove materials from its surface.

adhesion 1. The ability of the membrane to remain adhered during its service life to the substrate or to itself. 2. The ability of one material to remain secured to another material through the use of an adhesive.

adhesive A compound, glue, or mastic used in the application of gypsum board to framing or for laminating one or more layers of gypsum boards.

aggregate Gravel or crushed rock that, when mixed with sand, portland cement, and water, forms concrete.

amp (A) A measurement of the amount of electrical current in a circuit at any moment. *See* volt and watt.

asphalt A dark brown or black, solid or semisolid hydrocarbon produced from the residuum left after the distillation of petroleum, used as the waterproofing agent of a built-up roof and to impregnate roofing felt or fiberglass in the manufacture of shingles. It comes in a wide range of viscosities and softening points, from about 135°F (dead level asphalt) to 225°F (very step).

backfill The soil used to fill in an excavation next to a wall. This soil adds stability to the wall and keeps water away from it

bar joist Open-web, flat truss structural member used to support floor or roof structure. Web section is made from bar or rod stock, and chords are usually fabricated from "T" or angle sections.

batter boards A board frame supported by stakes set back from the corners of a structure that allows for relocating certain points after excavation. Saw kerfs in the boards indicate the location of the edges of the footings and the structure being built.

beam Load-bearing member spanning a distance between supports.

bearing Support area upon which something rests, such as the point on bearing walls where the weight of the floor joist or roof rafter bears.

bending Bowing of a member that results when a load or loads are applied laterally between supports.

bond 1. Any one of several patterns in which masonry units can be arranged. 2. To join two or more masonry units with mortar.

brick veneer Non-load-bearing brick facing applied to a wall to give the appearance of solid brick construction. Bricks are fastened to the backup structure with metal ties embedded in mortar joints.

bridging Members attached between floor joists to distribute concentrated loads over more than one joist and to prevent rotation of the joist. Solid bridging consists of joist-depth lumber installed perpendicular to and between the joists. Cross- bridging consists of pairs of braces set in an "X" form between joists.

butt joint Joints formed by the mill cut ends or by job cuts without a tapered edge. Syn-end joint.

cable Two or more insulated conductors wrapped in metal or plastic sheathing.

casement Glazed sash or frame hung to open like a door.

casing The trim around windows, doors, columns, or piers.

chalk line Straight working line made by snapping a chalked cord stretched between two points, transferring chalk to work surface.

circuit The path of electrical flow from a power source through an outlet and back to ground.

cladding Gypsum panels, gypsum bases, gypsum sheathing, cement board, etc., applied to framing.

codes Local laws governing safe wiring practices.

column Vertical load-bearing member.

conductors A wire or anything else capable of carrying electrical energy.

conduit Rigid or flexible tubing through which you can run wires.

control joint A groove tooled into a concrete slab during finishing to prevent uncontrolled cracking later on. These joints, to be effective, should be one-fourth the thickness of the slab.

corner brace Structural framing member used to resist diagonal loads that cause racking of walls and panels due to wind and seismic forces. May consist of a panel or diaphragm, or diagonal flat strap or rod. Bracing must function in both tension and compression. If brace only performs in tension, two diagonal tension members must be employed in opposing directions as "X" bracing.

course A row of masonry units. Most projects consist of several courses laid atop each other and separated by mortar.

coverage Area usually measured in square footage a given material will cover (i.e., 10 gallons per 1000 square feet of wallboard). Syn mileage distance.

dead level Absolutely horizontal, of zero slope.

deflection Displacement that occurs when a load is applied to a member or assembly. The dead load of the member or assembly itself causes some deflection as may occur in roofs or floors at mid-span. Under applied wind loads, maximum deflection occurs at mid-height in partitions and walls.

design load Combination of weight and other applied forces for which a building or part of a building is designed. Based on the worst possible combination of loads.

drain-waste-vent (DWV) system The network of pipes and fittings that carries liquid and solid wastes out of a building and to a public sewer, a septic tank, or a cesspool.

drywall Generic term for interior surfacing material, such as gypsum panels, applied to framing using dry construction methods, e.g., mechanical fasteners or adhesive.

edge venting The practice of providing regularly spaced openings at a roof perimeter to relieve the pressure of water vapor entrapped in the insulation.

elbow A fitting used to change the direction of a water supply line. Also known as an ell. Bends do the same thing with drain-waste-vent lines.

electrons Invisible particles of charged matter, moving at the speed of light through an electrical circuit.

expansion joint A structural separation between two building elements designed to minimize the effect of the stresses and movements of a building's components and to prevent these stresses from damaging the structure.

fire resistance Relative term, used with a numerical rating or modifying adjective to indicate the extent to which a material or structure resists the effect of fire.

fishing The process of getting cables through finished walls and ceilings.

fitting Any connector (except a valve) that lets you join pipes of similar or dissimilar size or material in straight runs or at an angle.

fixture 1. Any light or other electrical device that is permanently attached to a home's wiring. 2. Any of several devices that provide a supply of water or sanitary disposal of liquid or solid wastes. Tubs, showers, sinks, lavatories, and toilets are typical examples.

furring The act of applying furring strips to provide and air space between structural walls and the interior finish or to level an uneven surface.

general-purpose circuit Serves a number of light and/or receptacle outlets. *Also see* small-appliance and heavy-duty circuits.

glass foam Heat- and solvent-resistant, high compressive strength, friable, and fire- and water-resistant.

gravel stop Flanged device, normally metallic, designed to prevent loose aggregate from washing off the roof and to provide a finished edge detail for the built-up roof.

ground Refers to the fact that electricity always seeks the shortest possible path to the earth. Neutral wires carry it to ground in all circuits. An additional grounding wire, or the sheathing of metal-clad cable or conduit, protects against shock from a malfunctioning tool or other device.

header Horizontal framing member across the ends of the joists Also the member over a door or window opening in a wall.

hot wire The conductor that carries current to a receptacle or other outlet. *Also see* neutral wire and ground.

jamb One of the finished upright sides of a door or window frame.

joist Small beam that supports part of the floor, ceiling, or roof of a building.

junction box An enclosure used for splitting circuits into different branches. In a junction box, wires connect only to each other, never to a switch, receptacle, or fixture.

lintel Horizontal member spanning an opening such as a window or door. Also referred to as a header.

load Force provided by weight, external or environmental sources such as wind, water and temperature, or other sources of energy.

nominal Term indicating that the full measurement is not used; usually slightly less than the full net measurement, as with 2-×-4-inch studs that have an actual size when dry of 1½ × 3½ inches.

outlet Any potential point of use in a circuit. Receptacles, switches, and light fixtures are all considered outlets.

overhang The projection of a floor or roof over an outside wall.

pillar Column supporting a structure.

plate The top plate is the horizontal member fastened to the top of the studs or wall on which the rafters, joists or trusses rest; the sole plate is positioned at the bottom of the studs or wall.

plumb A term used to describe a true vertical line. If something is "out of plumb," it is not exactly straight up and down.

roll roofing Coated felts, either smooth or mineral-surfaced.

rough framing Structural elements of a building or the process of assembling elements to form a supporting structure where finish appearance is not critical.

rough-in The early stages of a construction project during which plumbing and electrical lines are run to their destinations. All work done after the rough-in, setting the fixtures and so forth, is finish work.

run Any length of pipe or pipes and fittings going in a straight line.

septic tank A reservoir that collects and separates liquid and solid wastes, then digests the organic material and passes the liquid waste onto a drainage field or a seepage pit. It is a private system counterpart of a municipal sanitary sewer and treatment plant.

sheathing Plywood, gypsum, wood fiber, expanded plastic or composition boards encasing walls, ceilings, floors and roofs of framed buildings. May be structural or non-structural, thermal-insulating or non-insulating, fire-resistant or combustible.

short circuit A condition that occurs when hot and neutral wires contact each other. Fuses and breakers protect against fire, which can result from a short.

sill Horizontal member at the bottom of door or window frames to provide support and closure.

spalling Fragmenting or breaking up of the surface, due to freeze-thaw reaction within the wall.

span Distance between supports, usually a beam or joist.

stop Strip of wood fastened to the jambs and head of a door or window frame against which the door or window closes.

stress Unit resistance of a body to an outside force that tends to deform the body by tension, compression or shear.

stud Vertical load-bearing or non-load-bearing framing member.

substrate Underlying material to which a finish is applied or by which it is supported.

tensile strength The maximum force, per unit of cross section area, which a material can sustain when it elongates to rupture.

thermal conductivity (k) Heat energy measured in BTUs per hour transferred through a one inch thick one square foot area of homogeneous material per degrees F. temperature difference from surface to surface.

vapor retarder Material used to retard the flow of water vapor through walls and other spaces where this vapor may condense at a lower temperature.

vent The vertical or sloping horizontal portion of a drain line that permits sewer gases to rise out of the house.

vermiculite An expanded aggregate used in lightweight insulating concrete.

volt (V) A measure of electrical potential. Volts × amps = watts.

water supply system The network of pipes and fittings that transports water under pressure to fixtures and other water-using equipment and appliances.

wet wall A strategically placed cavity (usually a 2-×-6 wall) in which the main drain/vent stack and a cluster of supply and drain-waste-vent lines are housed.

wood fiber board Heat- and solvent-resistant, fair "R" factor, good peel resistance, poor fire and water resistance.

Index

Illustrations are indicated by **boldface.**

D

I

S

T

Z

About the author

James Gerhart is experienced in residential construction, working for a number of years as a project coordinator. He built his reputation, however, teaching subjects such as math, estimating, scheduling, blueprint reading, and surveying to vocational and construction management studies students.

CPSIA information can be obtained
at www.ICGtesting.com
Printed in the USA
LVOW04s2034130516
488158LV00009B/132/P